Objective Floriculture and Landscape Architecture

(For JRF, SRF, DDA and ASRB NET)

Title	Author
Basic Concepts of Vegetable Science	*S.K. Tyagi*
Bonsai: An Art	*D.B. Singh*
Disease Problems in Vegetable Production 2nd Ed	*Satish K. Gupta*
Diseases of Vegetable Ornamental and Spice Crops	*Satish K. Gupta*
Economics in Horticulture	*M.J. Sargent*
Fruit and Vegetable Processing	*FAO*
Fundamentals of Vegetable Crop Production	*Beena Nair*
Hort-Agro Project Reports	*R.K. Sud*
Horticultural Engineering Technology Field Machinery	*R.C. Balls*
Horticulture and Livelihood Security	*Prem Nath*
Horticulture Training Manual	*DED LAOS*
Krishi Bagwani: Yojna Evam Prabandhan (Hindi)	*C.K. Bajpai*
Masala Faslo Ki Vaigyanik Kheti (Hindi)	*Kamini Kumari*
Organic Farming for Sustainable Horticulture	*P.P. Reddy*
Ornamental Plants for Gardening	*V.L. Chopra*
Plant Pathogens and Their Control in Horticulture	*G.R. Dixon*
Plant Protection in Horticulture (Vol. 1-3)	*P.P. Reddy*
Post Harvest Management and Production of Important Horticultural Crops	*Amit Nath*
Postharvest Technology of Kinnow	*A.D. Sharma*
Production Technology of Fruit Crops in Wasteland	*S.P. Singh*
Productivity Enhancing Technologies for Horticultural Crops	*P.P. Reddy*
Sabjiyo Ki Utpadan Prodhyogiki (Hindi)	*S.K. Arora*
Small-Scale Postharvest Handling Practices: A Manual for Horticultural Crops 4th Ed	*Lisa Kitinoja*
Spice Crops Production Technology	*P.C. Das*
Susk Bagwani Ky Sidhant (Principles of Arid Horticulture)	*M.L. Meena*
Sustainable Horticulture Development and Nutrition Security (Vol. 3)	*Prem Nath*
Systematic Pomology (Vol. 1-2) (Set) (Economical Edition)	*O.P. Pareek*
Vegetable Crops at a Glance	*S.K. Tyagi*
Vegetable Gums and Resins	*F.N. Howes*
Vegetables and their Allied as Protective Food	*M.K. Rana*

Objective Floriculture and Landscape Architecture

(For JRF, SRF, DDA and ASRB NET)

Kommu Pavan Kumar

Ph.D Research scholar

Dept. of Floriculture and Landscape Architecture

Bidhan Chandra Krishi Viswavidhyalaya,

Mohanpur, West Bengal

Heera Lal Atal

Ph.D Research scholar

Dept. of Floriculture and Landscape Architecture

Bidhan Chandra Krishi Viswavidhyalaya,

Mohanpur, West Bengal

SCIENTIFIC PUBLISHERS

Published by

SCIENTIFIC PUBLISHERS (INDIA)
5 A, New Pali Road, P.O. Box 91
Jodhpur 342 001 (INDIA)
E-mail: info@scientificpub.com
Website: http://www.scientificpubonline.com

ISBN: 978-93-90749-79-9 (PB)
ISBN: 978-93-90749-84-3 (EBook)

Printed in India

PREFACE

Floriculture is a fast emerging major venture in the world, especially as a potential money-spinner for many countries in the world.Floriculture is one of the most potent components of the Horticulture Industry, being important from aesthetic, social, and economic points of view. It has the potential for generating employment opportunities round the year and earning foreign exchange. In many countries, different floricultural value-added products are the main export items from the agriculture sector. During our masters and preparation for SRF, and ASRB/NET we feel that there is a lack of floriculture-related objective books, so we decided to write this type of book. This book is designed to give students ample opportunities to practice the objectives.This book is mainly aimed at graduate andpost-graduate students in agriculture and Horticulture.Thisbook has been developed to impart knowledge.This objective book on floriculture and landscaping architecture covering on various aspects of floriculture viz., importance and scope of floriculture and landscaping, history of gardening in India and styles and types of gardens, principles, and elements of landscape gardening, annual and perennial ornamental crops, principles and methods of propagation and production of ornamental crops. Most of the information has been taken from 'Ornamental Horticulture in India (G.S. Randhawa) and 'Glaustas Horticulture' (P. Muthukumar and R. Selvakumar). We hope this book will be a valuable and handy source of manual for amateur and professional students.Suggestions are invited from intellectuals, authors, and teachers. Despite all efforts, if any some errors may have persisted in the present text. The suggestions of readers are welcomed in this aspect too. Your suggestions will go a long way in the betterment of the present text. Any further suggestions for improving this book are welcome.

We heartly acknowledge all referees and our Doctoral degree classmates for their guidance and support.

Kommu Pavan Kumar

Heera Lal Atal

CONTENTS

MULTIPLE CHOICE QUESTIONS

1. Which one is the number one foliage plant in the world?
   (A) Dieffenbachia (B) Asparagus
   (C) Hedera (D) Helichrysum
2. Which one is known as foliage capital of the world?
   (A) California (B) The Netherlands
   (C) Apokka (USA) (D) Germany
3. Beautiful garden book is written by?
   (A) Vishnu Swarup (B) M.S. Randhawa
   (C) J.S. Arora (D) None of these
4. Which one is known as king of annual flowers?
   (A) Phlox (B) Salvia
   (C) Pansy (D) Marigold
5. Which is not primary colour?
   (A) Red (B) Orange
   (C) Yellow (D) Blue
6. Japanese flower arrangement is known as?
   (A) Jiyubana (B) Morimona
   (C) Moribana (D) Ikebana
7. Leading Jasmine perfume producing country?
   (A) France (B) Egypt
   (C) Bulgaria (D) The Netherland
8. Which is a blue colour variety of Rose?
   (A) Abhisharka (B) Priya
   (C) Samba (D) Pusa Christina
9. First pinching in Carnation is done at the stage of?
   (A) 4TH pair leaf (B) 8TH pair leaf
   (C) 6TH pair leaf (D) All of above
10. Petal bum in Chrysanthemum is due to deficiency of?
    (A) Boron (B) Calcium
    (C) Cobalt (D) Nitrogen
11. Nugget is a triploid variety of which flower?
    (A) Rose (B) Carnation
    (C) Marigold (D) Gladiolus

12. Exalbuminous seeds are found in which flower?
   (A) China aster (B) Orchid
   (C) Marigold (D) Tuberose
13. Which one is not a principle of landscaping?
   (A) Scale (B) Unity
   (C) Line (D) Balance
14. Overall effect of various features is known as?
   (A) Harmony (B) Unity
   (C) Mobility (D) Accent
15. How much area is classified for trees in garden?
   (A) 20-25 % (B) 15-20 %
   (C) 25-30 % (D) 20-30 %
16. Punjabi folk dance is an example of?
   (A) Harmony (B) Rhythm
   (C) Unity (D) Scale
17. Dropping growth habit is found in which plant?
   (A) *Salix babylonica* (B) *Rinus poptula*
   (C) Bottle brush (D) All of Above
18. In home garden lawn are should not be more than?
   (A) 15 % (B) 20 %
   (C) 30 % (D) 40 %
19. Germination % of *Ficus elastica*?
   (A) 55 % (B) 34 %
   (C) 70 % (D) 60 %
20. Planting time for deciduous ornamental plants?
   (A) April – June (B) Dec- Jan
   (C) Sept – Oct (D) All of Above
21. *Ficus benjamina* is scientific name of which tree?
   (A) Banyan tree (B) Pipal
   (C) Java fig tree (D) Rubber plant
22. *Salix babylonica* is a scientific name of which tree?
   (A) Poplar (B) Silver Oak
   (C) Weeping Willow (D) Devils tree
23. In double faced shrubbery tall shrubs are planted at?
   (A) Front (B) Back
   (C) Centre (D) None of these

24. Which annuals are suitable for screening propose?
    (A) Kochia (B) Hollyhock
    (C) Ipomoea (D) All of these
25. Which of the following is a day natural annual?
    (A) Gomphrena (B) Amaranths
    (C) Petunia (D) Cosmos
26. Which annual germinate only in dark?
    (A) Echium (B) Nigella
    (C) Nicotiana (D) All of these
27. Which warm colour dominated in winter?
    (A) Yellow (B) Red
    (C) Orange (D) All of these
28. Leading flowers seed producing state in India?
    (A) Haryana (B) Tamil Nadu
    (C) West Bengal (D) Punjab
29. Portulaca cultivar Jhumka is developed at?
    (A) IIHR (B) NBRI
    (C) PAU (D) IARI
30. Yellow is contrasting to which colour?
    (A) Orange (B) Violet
    (C) Blue (D) Red
31. Pseudo bulbs are found in which flower crop?
    (A) Tuberose (B) Canna
    (C) Orchid (D) All of these
32. Gladiolus is propagated by?
    (A) Bulb (B) Tuber
    (C) Corm (D) Rhizome
33. Swollen end of stolon is known as?
    (A) Tuber (B) Bulbils
    (C) Cormel (D) None of these
34. Gladiolus cultivar Shobha is developed through which method?
    (A) Hybridization (B) Selection
    (C) Mutation (D) None of these
35. Dahlia is propagated through?
    (A) Tuber (B) Rhizome
    (C) Cuttings (D) Tuberous root

36. Swami vinyanand was related to which flower crop?
(A) Dahlia (B) Gladiolus
(C) Chrysanthemum (D) Hibiscus

37. Scoring is important practice of which flower crop?
(A) Tuberose (B) Gladiolus
(C) Dahlia (D) All of these

38. Planting time for spring flowering crop?
(A) January (B) December
(C) November (D) August

39. Mayur variety of Gladiolus is released from?
(A) IIHR (B) NBRI
(C) IARI (D) AU Punjab

40. Which of the following is national flower of Italy?
(A) Lilium (B) Tulip
(C) Lily (D) Tuberose

41. How many chromosome numbers are there in Gladiolus?
(A) 30 (B) 40
(C) 60 (D) 16

42. Negative geotropism in Gladiolus is occur during?
(A) Vegetative period (B) Storage
(C) Transporting (D) Flowering

43. Breaking virus in tulip is transferred through?
(A) White fly (B) Thrips
(C) Aphids (D) Honey bee

44. Which of the following is dicotyledonous bulbs plant?
(A) Gladiolus (B) Oxalis
(C) Orchids (D) All of these

45. *Acroclinum roseam* is a botanical name of which flower?
(A) Dog flower (B) Duck flower
(C) Paper flower (D) None of these

46. Which of the following is does not have modified organs to climb?
(A) Climbers (B) Crippers
(C) Twiners (D) Both B & C

47. *Dracaena reflexa* is botanical name of which plant?
(A) Acalypha (B) Poinsettia
(C) Song of India (D) Dancing lady

48. Rolling of lawn is done to?
    (A) Levelling of ground (B) Touch nodes with ground
    (C) Cut the grass (D) All of these
49. Which of the following lawn grass does not require moving?
    (A) Bermuda grass (B) Zoysia grass
    (C) Agrostis grass (D) None of these
50. Fairy ring spot in lawn grass is due to?
    (A) Virus (B) Nutrient deficiency
    (C) Fungus (D) Bacteria
51. International registration authority of Rose is situated at?
    (A) USA (B) China
    (C) New Delhi (D) Bulgaria
52. Which state produce highest cut flower in India?
    (A) Karnataka (B) Tamil Nadu
    (C) Rajasthan (D) West Bengal
53. Hogarth course is known as?
    (A) Line of purity (B) Line of beauty
    (C) Line of trees (D) All of these
54. Tulip garden is situated at?
    (A) Srinagar (J & K) (B) Lucknow
    (C) Agra (D) Lahore
55. Which of the following flower is symbol of luxury?
    (A) Carnation (B) Stock
    (C) Statice (D) Lily
56. Which of the following tree is related to Buddha?
    (A) Ashoka (B) Sal
    (C) Palash (D) All of these
57. Pinjor garden Haryana is developed by?
    (A) King Hayder Ali (B) Fadai Khan
    (C) Bhuphindra Singh (D) Akbar
58. Poet Kali Das mentioned which plant in play Shakuntala?
    (A) Ashoka (B) Pipal
    (C) Madhvi (D) Kadam tree
59. Who was Le Notre?
    (A) Architecture (B) English garden designer
    (C) French garden designer (D) All of these

60. Which of following annual is suitable for screening propose?
(A) Hollyhock (B) Sweet pea
(C) Phlox (D) Both A & B

61. Which of following colour is dominant throughout the year in garden?
(A) Green (B) Red
(C) White (D) Grey

62. Gomphrena belongs to which family?
(A) Umbeliferae (B) Amaranthaceae
(C) Rutaceae (D) Astraceae

63. Which of the following is the cheapest method of lawn making?
(A) Dibbling of roots (B) Turfing
(C) Bricking (D) Seeding

64. Which of the following flower is known as barbeton daisy?
(A) Gladiolus (B) Gerbera
(C) Tulip (D) Pansy

65. Fluoride injury mostly found in which flower crop?
(A) Tuberose (B) Gladiolus
(C) Rose (D) All of these

66. Tinting is an important practice of which flower crop?
(A) Rose (B) Carnation
(C) Hibiscus (D) Jasmine

67. Which of the following bulbous crop cover highest area in India?
(A) Tuberose (B) Dahlia
(C) Gladiolus (D) Canna

68. Which of the following is a thorn less variety of Rose?
(A) Suchitra (B) Grand Gala
(C) Pusa Mohit (D) All of these

69. African Marigold is a symbol of?
(A) Vulgar mind (B) Jealousy and Sorrow
(C) Self-Steam (D) Purity

70. Who was developed herbaceous border?
(A) G. Jekwell (B) W. Robinson
(C) BK. Roy (D) Le Notre

71. Agave is used for which type of hedge?
(A) Dwarf ornamental (B) Tall ornamental
(C) Dwarf protective (D) Both A & B

72. Mahua tree belong to which family?
(A) Acantheceae (B) Anacardiaceae
(C) Sapotaceae (D) Rutaceae

73. Which colour flower found in *Bauhinia alba*?
(A) Purple (B) White
(C) Red (D) Yellow

74. Which type of climbing organs present in Bougainvillea?
(A) Tendrils (B) Thorns
(C) Rootlets (D) None of these

75. English flower arrangement is known as?
(A) Moribana (B) Morimona
(C) Jiyubana (D) Both A & B

76. Optimum spacing for hybrid tea Rose?
(A) 60 cm (B) 90 cm
(C) 75 cm (D) 65 cm

77. Which of the following is a thorn less species of Rose?
(A) *R. pendulina* (B) *R. bourboriana*
(C) *R. canina* (D) Both A & B

78. Which of the following is a multi-coloured mutant variety of Rose?
(A) Abhisharka (B) Madhos
(C) Priya (D) All of these

79. Hilling is an important operation of which flower crop?
(A) Tuberose (B) Chrysanthemum
(C) Carnation (D) Gladiolus

80. Which of the following Carnation developed through cross between *Dianthus caryophyllus* X *Dianthus chinensis*
(A) Marguerite (B) Perpetual
(C) Royal (D) None of these

81. Haldi Ghati is a variety of which flower?
(A) Rose (B) Dahlia
(C) Chrysanthemum (D) Carnation

82. Which of the following flower is known as Rose of Indies?
(A) Hibiscus (B) China aster
(C) Marigold (D) Dahlia

83. China aster is propagated through?
(A) Cuttings (B) Micro propagation
(C) Seeds (D) Rhizome

84. Which of the following orchid is monopodial?
(A) Dendrobium (B) Vanda
(C) Cattleya (D) Bulbophyllum

85. Which type of fruits are present in Rose?
(A) Capsule (B) Berry
(C) Nut (D) None of these

86. Maximum oil recovery is done by which species?
(A) J. sambac (B) J. Grandiflorum
(C) J. auriculatum (D) J. multiflorum

87. Mickey mouse is a variety of which flower?
(A) Carnation (B) Orchid
(C) Anthurium (D) Portulca

88. Chrysanthemum cultivar Money Maker is developed through?
(A) Biotechnology (B) Selection
(C) Hybridization (D) Mutation

89. Which of the following first variety of Floribunda Rose?
(A) La France (B) Rodhatte
(C) Buccaner (D) Grand gala

90. How much seeds of Marigold are required for one hectare?
(A) 1-1.5 Kg (B) 4-5 Kg
(C) 500 gm (D) 300 gm

91. Which of the following type SI found in Marigold?
(A) Protandary (B) Protogyani
(C) Both A & B (D) None of these

92. Which of the following largest genera of Orchid?
(A) Bulbophyllum (B) Dendrobium
(C) Vanda (D) Oncidium

93. *Zoysia tennifolia* is a botanical of which lawn grass?
(A) Manilla grass (B) Korean grass
(C) Korean velvet grass (D) Carpet grass

94. Which of the following storage disorder of Rose?
(A) Bull head (B) Bent neck
(C) Sleepiness (D) None of these

95. Which flower is the largest flower of the world?
(A) Lotus (B) Arun Titan
(C) Gurhal (D) Delphinium

96. Which of the following Rose variety have great demand in cut flower market?
(A) Raktagandha (B) Sindhoor
(C) Arjun (D) All of these

97. Big sized bottles with narrow mouth are used for house plants is known as?
(A) Aquarium (B) Terrarium
(C) Pots (D) Canes

98. Which of the following annual suitable for hanging baskets?
(A) Verbena (B) Petunia
(C) Zebrina (D) All of these

99. Arrangement of potted colourful plants around the plant is known as?
(A) Topiary (B) Trophy
(C) Rhythm (D) Carpet bedding

100. Who was introduced Mughal garden in India?
(A) Akbar (B) Babar
(C) Noorjhan (D) Hyder Ali

101. White Kachnar represent the which things?
(A) Extra Ordinary (B) Youth & Life
(C) Self-respect (D) All of these

102. Rashtrapti Bhavan garden is which style of garden?
(A) Japanese (B) English
(C) Mughal (D) None of these

103. Which of the following is not a type of Japanese garden?
(A) Hill garden (B) Flat garden
(C) Tea garden (D) Forcing garden

104. Who was coined term Bio aesthetic planning?
(A) Prof. Lancelot Hogben (B) W. Robinson
(C) Le Notre (D) None of these

105. Optimum size for Bonsai?
(A) 20-30 cm (B) 30-50 cm
(C) 30-60 cm (D) 50-60 cm

106. Floral clock is a special feature of which garden?
(A) Pinjor garden (B) Rock garden
(C) Lal Bagh garden (D) All of these

107. Optimum drying temperature for Chrysanthemum?
(A) 45-49 °C for 72 hours (B) 45-49 °C for 48 hours
(C) 45-49 °C for 96 hours (D) 45-49 °C for 25 hours

108. Which of the following is an off season variety of Tuberose?

(A) Rajat Rekha (B) Shringar

(C) Suvashini (D) Pearl Double

109. Nugget variety of Marigold was developed from?

(A) Egypt (B) USA

(C) India (D) Netherland

110. Rose variety Mohini in India introduce from?

(A) USA (B) China

(C) Germany (D) Italy

111. Inflorescence type of Chrysanthemum?

(A) Spike (B) Spadix

(C) Capitulum (D) None of these

112. What is mean of pot mums?

(A) 1 Plant in 1 Pot (B) 1 Species in 1 Pot

(C) 1 Species in >1 Pot (D) 1 Plant in >1 Pot

113. Who was introduce Damask Rose in India?

(A) Baber (B) British

(C) Araya (D) B.P. Pal

114. Scoring is an important operation of which flower?

(A) Tulip (B) Gladiolus

(C) Hibiscus (D) Rose

115. Hira-Niwa is the Japanese mean of which type Japanese garden?

(A) Sand garden (B) Flat garden

(C) Tea garden (D) Hill garden

116. Which of the following plant is suitable for Bonsai?

(A) *Ficus religiosa* (B) *F. benjamina*

(C) *F. bengalensis* (D) All of these

117. Which of the following is not a feature of English garden?

(A) Lawn (B) Running water

(C) Rockery (D) Herbaceous garden

118. Buddha Jayanti park is a which style of garden?

(A) Mughal (B) Japanese

(C) English (D) Persian

119. Which of the following is a best rooting media for Rose?

(A) Sand (B) Sand + FYM

(C) FYM + Coir (D) All of these

120. Foliage Nematode is a serious problem of which flower?
(A) Tuberose (B) Gladiolus
(C) Orchid (D) All of these

121. Which of the following type pollination found in Orchid?
(A) Animophilous (B) Hydrophilous
(C) Entomophilous (D) Ornithophilous

122. Economic part of Anthurium?
(A) Leaves (B) Spathe
(C) Whole plant (D) Spadix

123. Which of the following flower is an indicator of fluoride?
(A) Chandni (B) Gladiolus
(C) Tuberose (D) Kamini

124. Disbudding in Dahlia is done at which stage?
(A) Bud appearance stage (B) Pea stage
(C) Bud opening stage (D) Flower opening stage

125. Stem break is a post-harvest disorder of which flower?
(A) Dahlia (B) Gerbera
(C) Tuberose (D) Gladiolus

126. Lotus is propagated through which method?
(A) Suckers (B) Seeds
(C) Cuttings (D) All of these

127. Dr. B.P. Pal is a variety of which flower?
(A) Bougainvillea (B) Rose
(C) Chrysanthemum (D) Gerbera

128. Which type of pollination found in Heliconia?
(A) Ornithophilous (B) Self pollination
(C) Entomophilous (D) Animophilous

129. Which type of cross pollination mechanism found in Primula?
(A) Self Incompatibility (B) Dicogamy
(C) Heterostyle (D) Male sterility

130. Which of the following annuals is/are Indian origin?
(A) Gomphrena (B) Balsam
(C) Lady's lack (D) All of these

131. *Portulaca grandiflora* is a botanical name of which annual?
(A) Sun plant (B) Portulaca
(C) Sun flower (D) Both A & B

132. Which of the following Rose is mainly used for gulkhand?
(A) Damask Rose (B) Chinese Rose
(C) Edward Rose (D) Dog Rose

133. Colour break in Anthurium spathe is due to deficiency of?
(A) Calcium (B) Boron
(C) Nitrogen (D) Phosphorus

134. Which of the following pollinating agent in Petunia?
(A) Honey bees (B) Air
(C) Water (D) Birds

135. Propagation in Anthurium is done through?
(A) Rhizome (B) Slips
(C) Suckers (D) Bulbs

136. Yellow colour is extracted from which Rose species?
(A) *R. foetida* (B) *R. canina*
(C) *R. galica* (D) All of these

137. Semal tree is related to?
(A) Shiva (B) Kali
(C) Krishna (D) Buddha

138. Kamini is belongs to which family?
(A) Sapotaceae (B) Rutaceae
(C) Proteceae (D) Oleaceae

139. Shri Bajrang Bahadur Singh is related to which flower?
(A) Gladiolus (B) Chrysanthemum
(C) Jasmine (D) Gerbera

140. Which of the following plant is/are suitable for indoor condition?
(A) Aglaonema (B) Spider plant
(C) White flag plant (D) All of these

141. Growing of trees for forest purpose is known as?
(A) Arboriculture (B) Silviculture
(C) Arboretum (D) All of these

142. Trees are established in large waste land is known as?
(A) Woodland planting (B) Landscaping
(C) Architecture (D) Silviculture

143. Most suitable lawn grass for home garden?
(A) Bermuda grass (B) Korean grass
(C) Korean velvet grass (D) All of these

144. Wilt of Rose is caused by?
(A) Virus (B) Viroid
(C) Fungus (D) Bacteria

145. Apply artificial colour in Carnation is known as?
(A) Pulsing (B) Tinting
(C) Precooling (D) Hilling

146. Madhosh variety of Rose is developed through?
(A) Introduction (B) Mutation
(C) Selection (D) Demonstration

147. Which of the following flower is symbol of purity?
(A) Lily (B) Stock
(C) Lilium (D) Daffodil

148. Arka Archana is a variety of which flower?
(A) Zinnia (B) Dahlia
(C) China aster (D) Calendula

149. Pollination in Heliconia is done through?
(A) Honey bees (B) Huming bird & bats
(C) Sparrow (D) Both A & B

150. Propagation in Bougainvillea done through?
(A) Semi hard wood cutting (B) Hard wood cutting
(C) Layering (D) Soft wood cutting

151. Required ideal temperature for indoor plants?
(A) 10-20 ºC (B) 15-21 ºC
(C) 20-25 ºC (D) 8-10 ºC

152. Bronze colour foliage in Chrysanthemum is due to deficiency of?
(A) Boron (B) Led
(C) Calcium (D) Potash

153. Topple & bud rot in Lily is due to deficiency of?
(A) Calcium (B) Aluminium
(C) Iron (D) Cobalt

154. Which country is the largest exporter of floriculture products?
(A) Germany (B) UK
(C) USA (D) Netherland

155. India's share in the global flower market?
(A) 1 % (B) 0.6 %
(C) 2 % (D) 0.8 %

156. Biggest international flower market is established at?
(A) Aalsmeer (B) London
(C) Singapore (D) Australia

157. In which country highest consumption of flower per capita?
(A) Australia (B) Netherlands
(C) Switzerland (D) USA

158. Maximum consumption of cut Rose is during which of following event?
(A) Mother's day (B) X- mas day
(C) Valentine's day (D) Farmer's day

159. Chrysanthemum in Hindi is known as?
(A) Guldaudi (B) Genda
(C) Gurhal (D) Gulab

160. Bharat Sundari is an important variety of which flower?
(A) Hibiscus (B) Marigold
(C) Tulip (D) Dahlia

161. Banjaran cultivar of Rose is related to which type of Rose?
(A) Hybrid tea (B) Tea Rose
(C) Floribundas (D) None of these

162. Queen Elizabeth is a cultivar of which flower?
(A) Chrysanthemum (B) Rose
(C) Carnation (D) Lily

163. Birbal Sahnii is a variety of?
(A) Chrysanthemum (B) Dahlia
(C) Marigold (D) Tulip

164. African Marigold is an?
(A) Triploid (B) Diploid
(C) Aneuploidy (D) Tetraploid

165. Mostly the cultivars of Chrysanthemum are?
(A) Short day (B) Long day
(C) Day neutral (D) Semi short day

166. Which of the following practice is followed for large sized flowers?
(A) Pinching (B) Dishooting
(C) Disbudding (D) All of these

167. Which of the following is a growth retardant?
(A) B – Nine (B) ABA
(C) 2,4 – D (D) All of these

168. Which of the following substance is used for breaking Gladiolus corm dormancy?
(A) ABA (B) GA3
(C) NAA (D) SADH

169. Canna is Propagated through?
(A) Tuber (B) Rhizome
(C) Cuttings (D) Tissue culture

170. Shoot tip culture is commonly followed in which flowers?
(A) Carnation (B) Chrysanthemum
(C) Orchid (D) All of these

171. Thimma is an important cultivar of?
(A) Jasmine (B) Marigold
(C) Bougainvillea (D) Chrysanthemum

172. Spanish Jasmine is known as ....
(A) *J. grandiflorum* (B) *J. sambac*
(C) *J. auriculatium* (D) *J. multiflorum*

173. IIHR is located at ....
(A) Mumbai (B) Bangalore
(C) Shimla (D) Gurgaon

174. Damping off is caused by?
(A) *Pythium* (B) *Rhizoctonia*
(C) *Phytophthora* (D) All of these

175. Scarification of seeds is done by ....
(A) Acid (B) Hot water
(C) High temperature (D) All of these

176. Seeds priming is a process of controlled ....
(A) Hydration (B) Dormancy
(C) Germination (D) Infection

177. Which of the following is a foliage annual?
(A) Kochia (B) Coleus
(C) Ornamental cabbage (D) All of these

178. Which of the following drying method is suitable for Helichrysum?
(A) Air drying (B) Vacuum drying
(C) Sand drying (D) Press drying

179. Which of the following is a double bracts cultivar of Bougainvillea?
(A) Mahara (B) Mohan
(C) Cherry blossom (D) All of these

180. Mohini is a which colour cultivar of Rose?
(A) Blue (B) Chocolate
(C) Red (D) Yellow

181. Most beautiful flower on this globe?
(A) Rose (B) Orchid
(C) Lotus (D) Gladiolus

182. Jewel Orchids are valued for their beautiful?
(A) Leaves (B) Flower
(C) Roots (D) Stigma

183. *Eria pusilla* Orchid has how much height?
(A) 1 cm (B) 30 cm
(C) 5 cm (D) 10 cm

184. Fruit of Rose is known as ....
(A) Jip (B) Hip
(C) Column (D) Bud spot

185. Best time for budding in Rose in North Indian plains?
(A) October (B) November
(C) January (D) August

186. Hilly areas are most suitable for growing quality ....flowers
(A) Annual (B) Bulbous
(C) Biennial (D) Perennial

187. Pusa Tara is a variety of?
(A) Coleus (B) Celosia
(C) Coreopsis (D) Marigold

188. Which is an inter specific hybrid of Bougainvillea?
(A) Begum Sikander (B) Wazi Ali Shah
(C) Chitra (D) All of these

189. Closely related colours are known as?
(A) Contrasting (B) Complimentary
(C) Analogous (D) All of these

190. Yellow is contrasting to which colour?
(A) Orange (B) Violet
(C) Green (D) Purple

191. How many isolation distance should be keep for often cross pollinated flowers?
(A) 10 – 20 M (B) 50 – 60 M
(C) 100 – 150 M (D) 200 – 300 M

192. Who was given the classical theory of hybrid vigour?
(A) Morel (B) Mendel
(C) Shull (D) Lysenko

193. Gladiolus is originated from......?
(A) America (B) South Africa
(C) India (D) Mexico

194. Aglaonema and Crossandra are originated from........
(A) Mexico (B) China
(C) India (D) Japan

195. Mohini is a .... cultivar of Rose
(A) Diploid (B) Triploid
(C) Ancuploid (D) Haploid

196. How many seeds are there in 1 gm seed of Petunia?
(A) 100 (B) 1000
(C) 10000 (D) 50000

197. Flower abnormality character Cinderella is commonly found in?
(A) Bougainvillea (B) Begonia
(C) Balsam (D) Gladiolus

198. Osaka garden is located at?
(A) London (B) Japan
(C) America (D) Argentina

199. Vatsyana did not describe which of the following garden?
(A) Pramadodyan (B) Nandanvana
(C) Vrikshavatika (D) Brindavan

200. Who was laid out Dilkhusha garden at Lahore?
(A) Sahanjhan (B) Babar
(C) Akbar (D) Jahangir

201. Moorish garden style was developed at?
(A) Japan (B) France
(C) England (D) Spain

202. Royal Botanic garden of Kew is located at?
(A) America (B) England
(C) Germany (D) Netherland

203. Bradari is an important feature of .... garden
(A) Japanese (B) English
(C) Mughal (D) Persian

204. Rashtrapti Bhavan garden is laid out by..........
(A) Queen Elizabeth (B) Sir Edwin Lutyen
(C) Le Notre (D) G. Jkwell

205. In which style of gardening first plan is made on paper and then apply on field?
(A) Informal (B) Formal
(C) Wild (D) Free

206. Mandoor garden Jodhpur is laid out by ....
(A) Abhai Singh (B) Vinay Singh
(C) Vijay Singh (D) Vikarm Singh

207. Pinjoor garden is laid out by...........
(A) Fadai khan (B) Abhijit Singh
(C) Akbar (D) Wazi Ali Shah

208. Brindavan garden was laid out near .... River
(A) Krishna (B) Kaveri
(C) Godavari (D) Yamuna

209. Stone Lanterns are important feature of .... Garden
(A) Mughal (B) Japanese
(C) English (D) Persian

210. Rose is national flower of ....
(A) UK (B) USA
(C) Iran (D) All of these

211. Optimum storage temperature for gladiolus corm is .... °C
(A) 0 – 1 °C (B) 2 – 5 °C
(C) 4 – 7 ° (D) 10 -12 °C

212. Which of the following is not a fragrant variety of Gladiolus?
(A) Lucky star (B) Jimmy boy
(C) Sager (D) All of these

213. Punjab gold is a variety of ....
(A) Marigold (B) Chrysanthemum
(C) Jasmine (D) Gladiolus

214. Storage corm rot in Gladiolus is caused by.........
(A) Fusarium (B) Pythium
(C) Panicillium (D) Rhizoctinia

215. Arthur Sim Carnation is highly resistance to ....
(A) Wilt (B) Rust
(C) Stem rot (D) All of these

216. Who was started tissue culture in Orchid?
(A) Lysenko (B) Shull
(C) Morel (D) S. Srivastava

217. Calyx splitting in Carnation is due to ....
(A) Genetic factor (B) Environmental factor
(C) Nutritional factor (D) All of these

218. Most serious pest of Carnation.........
(A) Aphids (B) Thrips
(C) Red spider mite (D) White fly

219. Gulkhand is prepared by using petals and sugar in ratio of ......
(A) 1:1 (B) 1:2
(C) 2:1 (D) 1:3

220. Chemical defoliation in Rose is done by ....
(A) Urea (B) $CuSo_4$
(C) GA3 (D) ABA

221. Which colour Roses are most popular?
(A) Red (B) Pink
(C) Yellow (D) Blue

222. In cut Roses minimum vase life should be .... days
(A) 10 (B) 12
(C) 8 (D) 14

223. Tuberose cultivar Suvasini is released from ....
(A) IARI (B) NBRI
(C) IIHR (D) BCKV

224. Dahlia tuberous roots should be stored in ....
(A) Cold storage (B) Sand
(C) Coir dust (D) CAS

225. Spotted wilt in Dahlia is caused by ....
(A) Fungus (B) Virus
(C) Bacteria (D) Genetic

226. Dr. B.P. Pal was developed how many cultivars of Rose?
(A) 100 (B) 102
(C) 200 (D) 105

227. Normal humidity in green house should be .... percent
(A) 30 – 40 (B) 20
(C) 60 – 65 (D) 70 - 80

228. Diploid Chromosome number in Gerbera........

(A) 30 (B) 50

(C) 60 (D) 18

229. Anthurium flower can be stored at .... temperature

(A) 2 °C (B) 13 °C

(C) 5 °C (D) 10 °C

230. Which is a serious disease of Anthurium?

(A) Rust (B) Wilt

(C) Anthracnose (D) All of these

231. *Hemerocallis fulva* is botanical name of ....

(A) Day Lily (B) Sword Lily

(C) Tiger Lily (D) All of these

232. Bird of Paradise is native to..........

(A) India (B) America

(C) South Africa (D) China

233. China aster is a .... crop

(A) Self-pollinated (B) Cross pollinated

(D) Often cross pollinated (D) None of these

234. Narcissus flowers are harvested at which stage?

(A) Tight bud stage (B) Goose neck stage

(C) Half open stage (D) Fully open stage

235. Pin and Thrum type flowers are found in which flower?

(A) Petunia (B) Primula

(C) Poinsettia (D) Orchid

236. Which one is fully compatible?

(A) $S_1 \times S_2 \times s_3 \times s_4$ (B) $S_1 \times S_2 \times S_3 \times S_4$

(C) $S_1 \times s_2 \times s_3 \times s_4$ (D) All of these

237. Which of the following male sterility is commonly found in Ornamental plants?

(A) CGMS (B) CMS

(C) GMS (D) All of these

238. Single gene male sterility is transferred in .... years

(A) 2–3 (B) 7–8

(C) 4-5 (D) 10

239. Virus free plants produced through...........

(A) Anther culture (B) Meristem culture

(C) Seed (D) Cuttings

240. Which aneuploids is mostly used in Ornamentals?
(A) Nullisomic (B) Monosomic
(C) Trisomic (D) Tetrasomic

241. Nullisomic means ....
(A) 2n + 1 (B) 2n + 2
(C) 2n – 1 (D) 2n – 2

242. Lilium bulbs are stored at............... ºC
(A) 2-3 (B) -2-2
(C) 8-10 (D) 7-8

243. Representation of the structure of any surface is known as ....
(A) Shape (B) Canopy
(C) Texture (D) Scale

244. Optimum water holding capacity of potting media should be .... percent
(A) 30–40 (B) 40-50
(C) 50-60 (D) 70-80

245. Soluble salt in potting media should be .... ppm
(A) <100 (B) >100
(C) <500 (D) >500

246. Which pots are most suitable for Cacti?
(A) Cement (B) Plastic
(C) Soil (D) All of these

247. In normal air how much Ethylene is present?
(A) 5-10 ppb (B) 3-5 ppb
(C) 3-5 ppm (D) 100-200 ppb

248. Bent neck in Rose is very common when grow at ....
(A) High light (B) Low light
(C) High temperature (D) Low temperature

249. HQC is an effective ....
(A) Germicide (B) Fungicide
(C) Bactericide (D) Weedicide

250. Salt of HQ is a most common ....
(A) Germicide (B) Fungicide
(C) Bactericide (D) Weedicide

251. Green Glo is used for improve longevity of.........
(A) Flower (B) Foliage
(C) Fruit (D) All of these

252. CO_2 concentration in normal air is about..........
(A) 200 ppm (B) 300 ppm
(C) 400 ppm (D) 500 ppm

253. Which of the following is a complex morphological unit?
(A) Fruit (B) Flower
(C) Vegetable (D) All of these

254. Water disinfection is done by ....
(A) Sodium chloride (B) Sodium hypo chloride
(C) Sodium Bai chloride (D) Potassium chloride

255. Bluing of Rose petals is due to ....
(A) Accumulation of Ammonia (B) Nitrogen deficiency
(C) Calcium deficiency (D) Boron deficiency

256. Hypobaric storage is also known as ....
(A) Modified atmospheric storage (B) Controlled atmospheric storage
(C) Low pressure storage (D) Cold storage

257. Limp neck is disorder of ....
(A) Rose (B) Tulip
(C) Dahlia (D) Gurhal

258. Quilling of florets is a common disorder of ....
(A) Tuberose (B) Gladiolus
(C) Chrysanthemum (D) Dahlia

259. Leaf scorching in Lily is due to .... deficiency
(A) Boron (B) Fluorine
(C) Manganese (D) Aluminium

260. Which site is appropriate for Lawn in home garden?
(A) South (B) South East
(C) South West (D) All of these

261. In one mowing of Lawn how much grass showed is cut?
(A) 1/2 (B) 1/3
(C) 1/4 (D) Any of the above

262. Fairy ring spot in Lawn is due to ....
(A) Nutrient deficiency (B) Fungus
(C) Bacteria (D) Virus

263. *Delonix regia* is commonly known as ....
(A) Blue Gulmohar (B) Red Gulmohar
(C) Yellow Gulmohar (D) All of these

264. Kigellia has beautiful ....

(A) Flower (B) Foliage

(C) Fruits (D) Bracts

265. Neem belong to family ....

(A) Bignoniaceae (B) Maliaceae

(C) Sapoteceae (D) Tilleaceae

266. Which is/are fragrant flower tree?

(A) Magnolia (B) Devil's tree

(C) Michelia (D) All of these

267. A shade tree should have .... canopy

(A) Wide oval (B) Wild

(D) Round (D) Pyramidal

268. Which is/are white colour flower tree?

(A) Magnolia (B) Mimusops

(C) Plumeria (D) All of these

269. Thuja is grown for ....

(A) Flower (B) Fruits

(C) Cut foliage (D) Edge

270. Which species of *Ficus* is grown for religious purpose?

(A) *bengalensis* (B) *elastic*

(C) *religiosa* (D) *carica*

271. *Terminalia* .... Have medicinal value

(A) *catapa* (B) *arjuna*

(C) *bellerica* (D) All of these

272. Bougainvillea is used for ....

(A) Climber (B) Shrubs

(C) Pot plant (D) All of these

273. *Passiflora caerulea* has .... shaped flower

(A) Watch (B) Butterfly

(C) Birds (D) Duck

274. *Aristolchia elegans* has .... shaped flower

(A) Duck (B) Birds

(C) Watch (D) Butterfly

275. *Pyrostegia venusta* has .... colour flower

(A) Red (B) Orange

(C) Pink (D) White

276. *Beaumontia grandiflora* is native to …………
(A) America (B) India
(C) China (D) South Africa

277. Duck flower is native to ………..
(A) North America (B) South America
(C) South Africa (D) India

278. A perennial plant having distinct trunk and crown at top is known as
(A) Shrub (B) Herb
(C) Climber (D) Tree

279. Which colour flower shrub is grown for night garden?
(A) Red (B) Yellow
(C) White (D) Green

280. Which of the following have beautiful bracts?
(A) Bougainvillea (B) Mussaenda
(C) Poinsettia (D) All of these

281. Which is the conservation of water through creative landscaping
(A) Ornamental Horticulture (B) Xeriscaping
(C) Landscaping (D) Water harvesting

282. *Acacia farnesiana* is an ……….
(A) Tall Ornamental (B) Dwarf Ornamental
(C) Tall protective (D) Dwarf Protective

283. Formal Edging is made with ……….
(A) Stone (B) Bricks
(C) Both A & B (D) Plant

284. *Punica grantum* has which colour flower?
(A) Pink (B) Red
(C) Orange (D) Purple

285. Hibiscus belong to family ………..
(A) Irrideceae (B) Aspragecea
(C) Malvaceae (D) Astraceae

286. *Codiaeum variegatum* performe very well at ………
(A) Pune (B) Banglore
(C) Delhi (D) Dargilling

287. China Shoe flower is known as ……….
(A) Hibiscus (B) Gurhal
(C) Both A & B (D) Dahlia

288. Most of Palms are propagated by ....
(A) Suckers (B) Seeds
(C) Cuttings (D) Slips

289. How many genera are included in Cycadaceae family?
(A) 8 (B) 9
(C) 15 (D) 20

290. Furcraeae is propagated through ....
(A) Cuttings (B) Bulbils
(C) Rhizome (D) Bulb

291. Which of the following is cabbage Rose?
(A) R. foitida (B) R. gallica
(C) R. centifolia (D) R. canina

292. High quality perfume is produced from...........
(A) J. sambac (B) J. multiflora
(C) J. grandiflorum (D) J. auriculatum

293. Sweet William is belong to family ....
(A) Rutaceae (B) Bracicaceae
(C) Caryophyllaceae (D) Astraceae

294. Stock is belong to family ....
(A) Bracicaceae (B) Araceae
(C) Plumbaginaceae (D) Nactaginaceae

295. Arka Ambra is a variety of ....
(A) Gladiolus (B) Crossandra
(C) Carnation (D) Dahlia

296. Pusa Basanti is a variety of ....
(A) Chrysanthemum (B) Marigold
(C) Dahlia (D) Petunia

297. Parrot flower is also known as ....
(A) Heliconia (B) Iris
(C) Sweet sultan (D) Sweet William

298. How much Rose petals required for 1 Kg Rose oil?
(A) 10 Ton (B) 3-4 Ton
(C) 10-15 Ton (D) 8-10 Ton

299. Precursor of Ethylene ....
(A) Methionine (B) Tetracycline
(C) Terthelion (D) None of these

300. Which of the following is a variety of Tuberose?
(A) PG purple (B) PG White
(C) Arka Nirantra (D) Arka Kamini

301. M.A. Kher was associated with ....
(A) Gladiolus (B) Chrysanthemum
(C) Marigold (D) Dahlia

302. Growing of Flowers in standing acid water is known as ....
(A) Bog garden (B) Rockery garden
(C) Flat garden (D) None of these

303. Which of the following Ornamental plant not native to India?
(A) Hedychum (B) Lonicera
(C) Jacaranda (D) Tabernaemontana

304. Miracle garden is located at.........
(A) London (B) Dubai
(C) Bagdad (D) England

305. Which of the following country have highest per capita consumption of cut flower?
(A) Switzerland (B) Netherland
(C) Germany (D) China

306. Who was the first Mughal empire introduced Mughal garden in India?
(A) Akbar (B) Babar
(C) Jahangir (D) Sahanjha

307. Following cultivar of Bougainvillea not developed through interspecific hybridization?
(A) Chitra (B) Wazi Ali Shah
(C) Matama Gandhi (D) Begaum Sikandar

308. In which type of garden the element of surprise is lost?
(A) Informal (B) Formal
(C) Persian (D) Wild garden

309. Which of the following is not a symmetrical garden?
(A) Rashtrapati Bhavan (B) Pinjor garden
(C) Taj Mahal (D) Budh Jayanti park

310. Which of the following is a deciduous tree?
(A) Juniperus (B) Criptomeria
(C) Salix (D) Magnolia

311. Identify a shrub
    (A) *Prunus serrulata* (B) *Morus alba*
    (C) Podocarpus (D) *Aucaba japonica*
312. The key feature of British garden
    (A) Herbaceous border (B) Baradari
    (C) Water (D) Stones
313. Identify the fragrant shrub
    (A) Tabarraemontuna (B) Polycias
    (C) Mussanda (D) Graptophyllum
314. Which plant is used for bedding purpose?
    (A) Pansy (B) Daisy
    (C) Sweet pea (D) Phlox
315. The plant which is suitable for grow in hanging basket?
    (A) Verbena (B) Alyseum
    (C) Nasturtium (D) All the above
316. Which one require partial sun light?
    (A) Pisonia (B) Polyscias
    (C) Hibiscus (D) Crotons
317. Which of the following is can grow in fully shade condition?
    (A) Eranthemum (B) Polyscias
    (C) Graptophyllum (D) Bougainvillea
318. Which climber is useful for screening purpose?
    (A) Porana (B) Bignonia
    (C) Scindapsus (D) Clitoria
319. Flowering climber which can grow in partial shade condition?
    (A) Aristolochia (B) Philodendron
    (C) Adenocalymma (D) Lonicera
320. Climber which one is suitable for pot culture?
    (A) Quisqualis (B) Tecoma
    (C) Tristellatia (D) Jasminum
321. Which climber is suitable for trellis for the purpose only as a light climber?
    (A) *Allamanda cathartica* (B) *Adenocalymma allicea*
    (C) *Petrea volubilis* (D) *Solanum seaforthianum*
322. Which of the following climber has foliage habitat?
    (A) *Monstera deliciosa* (B) *Passiflora edulis*
    (C) *Jasminum. Sp.* (D) *Jacquemontia violaceae*

323. Which of the following is suitable for grow as a edge purpose?
(A) Eupatorium (B) Alternenthera
(C) Duranta (D) All the above

324. The plant which is suitable for Topiary purpose?
(A) Cupressus (B) Tecoma
(C) Graptophyllum (D) Alyseum

325. Which tree is suitable for roof gardening?
(A) Canna (B) Anthirrhinum
(C) Mimusops (D) Vernonia

326. Plant suitable for water gardening
(A) Nuphar luteum (B) Nymphaea spp.
(C) Duck weed (D) All the above

327. Which plant is suitable for Bog gardens?
(A) Sagittaria (B) Cupressus
(C) Alternenthera (D) Pelargonium

328. The plant suitable for shade garden
(A) Digitalis (B) Candytuft
(C) Ageratum (D) Magnolia

329. Plant suitable for dish garden
(A) Snake plant (B) Azalea
(C) Salix (D) Camellia

330. Which one is white fragrant flower tree?
(A) *Casuarina equisetifolia* (B) *Grevilea robusta*
(C) *Millingtonia hortensis* (D) *Butea monosperma*

331. Which of the following is a winter annual?
(A) Balsam (B) Petunia
(C) Kochia (D) Tithonia

332. Which of the following is Cycad?
(A) Caryota urens (B) Zamia
(C) Pritchardia (D) Adiantum

333. The plant part which is modify into a spine
(A) Flowers (B) Roots
(C) Leaves (D) Stem

334. Rock plants need replanting in .... years
(A) 4-5 (B) 2-3
(C) 3-4 (D) 5-6

335. Polypodium is an …………

(A) Succulent (B) Cacti

(C) Fern (D) Winter annual

336. Agave belong to the family …………..

(A) Euphorbiaceae (B) Aspragusceae

(C) Amaryllidaceae (D) Amranthaceae

337. Suitable shed net for Codiaeum

(A) White (B) Red and Grey

(C) Green (D) All of above

338. *Alpinio speciose* is an …………

(A) Cacti (B) Succulent

(C) Bulbous plant (D) Fern

339. Beaumontia grandiflora produce ………… colour flowers

(A) Yellow (B) White

(C) Red (D) Pink

340. In which type of Bonsai have single trunk and twisted once or more times?

(A) Cascade (B) Oblique

(C) Upright (D) Winding

341. Mother Teresa is a variety of ……….

(A) Tuberose (B) Rose

(C) Chrysanthemum (D) Dahlia

342. Ornamental plant which is recommended for shade in town roads?

(A) Ferrugineum (B) Bauhinia

(C) Lagerstroemia (D) Stericulia

343. Which plant is suitable for planting beside highways?

(A) Mahua (B) Dalbergia

(C) Shorea (D) All the above

344. Which plant is associated with lord Krishna?

(A) *Saraca indica* (B) *Anthocephalus kadamba*

(C) *Ficus benghalensis* (D) *Bassia latifolia*

345. The following flowering tree which is recommended for planting along the banks of rivers and canals

(A) *Barringtonia acutingula* (B) *Delonix regia*

(C) *Peltophorum ferrugineum* (D) All of these

346. What is the medium range requirement of light for good growth of indoor plants?
(A) 1000-2000 lux (B) 2000-5000 lux
(C) 1500-5000 lux (D) < 750 lux

347. How much light quantity require for Boston fern good growth?
(A) 10000 lux (B) 1500-5000 lux
(C) 1000-2000 lux (D) 2000-5000 lux

348. Which of the following indoor plant require low light intensity for its growth?
(A) *Monestera deliciosa* (B) *Caryota mitis*
(C) *Pilea involucrata* (D) *Coleus blumei*

349. Which indoor plant require 1500-10000 lux light?
(A) *Hoya carnosa* (B) *Asparagus setaceus*
(C) *Aspidistra elatior* (D) *Ficus lyrata*

350. Which indoor plant require 1000 lux light?
(A) Snake plant (B) Gold dust plant
(C) Corn plant (D) Wax plant

351. The main aim of landscaping in industrial areas is to reduce
(A) Noise (B) Wind velocity
(C) Pollution (D) All of these

352. Which plant is suitable for interiorscaping?
(A) *Amherstia nobilis* (B) *Pithecolobium saman*
(C) *Eugenia operculata* (D) *Peperomia caperata*

353. The tree which is tolerant to SO_2?
(A) *Acacia nilotica* (B) *Ailanthus excels*
(C) *Thuja compacta* (D) *Pitheselobium dulce*

354. The tree which is tolerant to fluoride?
(A) Artocarpus (B) Eucalyptus
(C) Casuarina (D) Delonix

355. Which tree is suitable for cement factories?
(A) *Butea monosperma* (B) *Grevilllea robosta*
(C) *Holoptelea integrifolia* (D) All the above

356. Which tree is suitable for thermal power?
(A) *Lagerstroemia indica* (B) *Tamarindus indica*
(C) *Tectona grandis* (D) All of these

357. Which tree species is suitable for temples?
(A) *Plumeria alba* (B) *Ailanthus excels*
(C) *Pitheselobium dulce* (D) *Butea monosperma*

358. Which lawn grass is suitable for acid soil?
(A) *Sporobolus tremulus* (B) *Poa pratensis*
(C) *Cynodon dactylon* (D) *Stenotaphrum secundatum*

359. The lawn grass which is suitable for saline soil?
(A) Kikiyu grass (B) Japan grass
(C) Manila grass (D) Chain grass

360. Botanical name of Manilla grass
(A) *Zoysia japonica* (B) *Zoysia matrella*
(C) *Zoysia tenuifolia* (D) *Poa pratensis*

361. Dr. Foja Singh is associated with which flower?
(A) Carnation (B) Orchid
(C) Chrysanthemum (D) Jasmine

362. Which one multicolour mutant variety of rose?
(A) Madhosh (B) Priya
(C) Pusa Christina (D) Abhisarkha

363. Stem break in Gerbera is due to deficiency of ....
(A) Led (B) Calcium
(C) Boron (D) Nitrogen

364. At which stage Rose cut flower harvest for distant market?
(A) Cup stage (B) Colour shown
(C) Tight bud stage (D) Sepal reflect

365. Man in Japanese language is called as ....
(A) Shin (B) Soe
(C) Jushi (D) Heike

366. Hira-niwa is the Japanese mean of which type garden?
(A) Flat garden (B) Sand garden
(C) Passage garden (D) Tea garden

367. Chrysanthemum is also known as ....
(A) Glory of east (B) Queen of east
(C) Autumn queen (D) All of these

368. Which of the following bulbous crop have highest area in India?
(A) Tuberose (B) Canna
(C) Gladiolus (D) Lily

369. Tajmahal is a variety of which flower?
(A) Carnation (B) Rose
(C) Dahlia (D) Tulip

370. Which of the following is also known as moss Rose?
(A) Sweet William (B) Portulaca
(C) Chrysanthemum (D) Carnation

371. Which of the following plant is known as flame of forest?
(A) Arjun (B) Palas
(C) Tecoma (D) Siris

372. Bonsai is an Art of ....
(A) China (B) Japan
(C) France (D) America

373. Which of the following national flower of Netherland?
(A) Dahlia (B) Tulip
(C) Gerbera (D) Carnation

374. Which part of Rose plant is most susceptible to low temperature?
(A) Leaf (B) Bud union
(C) New shoot (D) Root

375. Carnation is also known as ....
(A) Dog flower (B) Devine flower
(C) Paper flower (D) Parrot flower

376. *Gypsophilla paniculata* is a botanical name of ....
(A) Baby's breath (B) Gompherena
(C) Gypsophilla (D) Both A & C

377. Statice is belong to family ....
(A) Araceae (B) Astraceae
(C) Plumbaginaceae (D) Bignoniaceae

378. Petunia is belong to which family?
(A) Oliaceae (B) Solaneceae
(C) Astraceae (D) Scrophuleraceae

379. Disco is a variety of ....
(A) Chrysanthemum (B) Marigold
(C) Carnation (D) Petunia

380. Canna is propagated by ....
(A) Tuber (B) Rhizome
(C) Bulb (D) Corm

381. Which of the following is the symbol of women's love?
(A) Rose (B) Daffodil
(C) Carnation (D) Stock

382. Peter Pears is a cultivar of ....
(A) Rose (B) Carnation
(C) Gladiolus (D) Orchid

383. Which of the following is not a short day plant?
(A) Poinsettia (B) Bryophyllum
(C) Viola (D) Spider plant

384. R.N. Bhatt was associated with which flower?
(A) Jasmine (B) Gladiolus
(C) Hibiscus (D) Rose

385. Which is not a component of beauty?
(A) Line (B) Shape
(C) Colour (D) Scale

386. Running water is the life of .... Garden
(A) Japanese (B) English
(C) Mughal (D) Persian

387. Bryant Park is located at
(A) Conoor (B) Kodaikanal
(C) Ooty (D) Kolkata

388. Which of the following is not a blue colour flower annual?
(A) Linaria (B) Corn flower
(C) Ageratum (D) Nigella

389. Which annual is not a hardy annual?
(A) Oxalis (B) Sweet pea
(C) Digitalis (D) Viola

390. Star flower is belong to family ....
(A) Asteraceae (B) Polemoniaceae
(C) Lamiaceae (D) Brassicaceae

391. An area of garden devoted exclusively to shrubs is called as ....
(A) Edge (B) Hedge
(C) Shrubbery border (D) Trophy

392. Which of the following is not a foliage shrub?
(A) Aralia (B) Pisonia
(C) Calliandra sp. (D) Acalypha

393. Dancing lady is belong to family.......
(A) Acanthaceae (B) Onagraceae
(C) Rubiaceae (D) Fabaceae

394. The term Topiary is derived from .... word
(A) French (B) Greek
(C) Roman (D) Latin

395. Kamala Nehru Park Mumbai is famous for...........
(A) Tulip (B) Rose
(C) Topiary (D) Floral clock

396. Who is the father of road side avenue planting?
(A) Buddha (B) Asoka
(C) Hyder Ali (D) Fadhai khan

397. Neem and Tamarind have ability to filter ....
(A) Noise (B) Dust
(C) Light (D) All of these

398. Which of the following is not a fast growing tree?
(A) Gulmohar (B) Neem
(C) Tamarind (D) Copper shield tree

399. Siris tree is belong to family
(A) Magnoliaceae (B) Fabaceae
(C) Myrtaceae (D) Meliaceae

400. Which of the following flower is known as pot Marigold?
(A) Carnation (B) Calendula
(C) Marigold (D) Hibiscus

401. Black Prince is a .... Colour variety of Gladiolus
(A) Red (B) Yellow
(C) White (D) Orange

402. Which variety is developed through cross between Black Jack X Friendship?
(A) Shobha (B) Priscilla
(C) Nazrana (D) Peter pears

403. Dahlia is originated from ....
(A) China (B) India
(C) Mexico (D) South Africa

404. Swami Vinayanda is related to ....
(A) Dahlia (B) Hibiscus
(C) Marigold (D) Gladiolus

405. R.Z. Lynch is associated with which flower?
(A) Gladiolus (B) Pot Marigold
(C) Gerbera (D) China aster

406. Gilly flower is a common name of ....

(A) Day lily (B) Stock

(C) Statice (D) Bougainvillea

407. Which of the following is a bicoloured variety of Bougainvillea?

(A) Thimma (B) Partha

(C) Usha (D) Sholay

408. Which of the following flower have Nematicide property?

(A) Marigold (B) Zinnia

(C) Hibiscus (D) Both A & B

409. Which of the following is not a Twiner?

(A) Echitis (B) Hiptage

(C) Morning Glory (D) Loniccra

410. Which is known as queen of climbers?

(A) Sky flower (B) Virgin flower

(C) Railway creeper (D) Duck flower

411. *Caryota urens* is a botanical name of ....

(A) Sago palm (B) Royal palm

(C) Fish tail palm (D) Butterfly palm

412. Cactus is originated from?

(A) South America (B) North America

(C) South Africa (D) Argentina

413. Cacti become lanky and do not flower when grow in ....

(A) High humidity (B) Shade

(C) Low temperature (D) Sunny situation

414. Which is not a tender bulbs plant?

(A) Gladiolus (B) Tuberose

(C) Daffodil (D) Dahlia

415. Which of the following house plant is suitable for terrarium?

(A) Hedera (B) Maranta

(C) Begonia (D) Aracaria

416. Ghost tree is belong to which family?

(A) Araceae (B) Araliaceae

(C) Asparagaceae (D) Lamiaceae

417. Annuals lost seed viability within ....

(A) 3-4 Months (B) 6-12 Months

(C) 4-6 Week (D) 10-12 Week

418. Crooked neck is physiological disorder of ....

(A) Tuberose (B) Chrysanthemum

(C) Rose (D) Dahlia

419. Garden word is originated from which language term?

(A) Greek (B) Latin

(C) French (D) Sanskrit

420. Jwahar Lal Nehru is a variety of ....

(A) Dahlia (B) Carnation

(C) Hibiscus (D) Bougainvillea

421. Division of Floriculture and Landscape at New Delhi was opened in ....

(A) 1971 (B) 1969

(C) 1983 (D) 1989

422. National registration centre for Chrysanthemum is located at ....

(A) FRI Dehradun (B) NBRI Lucknow

(C) IARI New Delhi (D) IIHR Bengaluru

423. Where is the highest per capita consumption of live plants?

(A) Australia (B) England

(C) Germany (D) Norway

424. 25th December is celebrated for ....

(A) Women's day (B) Christmas day

(C) Easter day (D) Children day

425. Which flower have maximum area in India?

(A) Gladiolus (B) Jasmine

(C) Marigold (D) Tuberose

426. Rose variety Mohini is belong to which species?

(A) Multiflora (B) Grandiflora

(C) Floribunda (D) Barboniana

427. Banjaran cultivar is belong to which group of Rose?

(A) Multiflora (B) Grandiflora

(C) Floribunda (D) Barboniana

428. Pusa Arpita is a variety of ....

(A) French Marigold (B) African Marigold

(C) Zinnia (D) Chrysanthemum

429. Tuberose Cultivar Rajat Rekha is developed through ....

(A) Hybridization (B) Selection

(C) Mutation (D) Introduction

430. Which of the following acids are following for gladiolus corm dormancy?
(A) Stearic (B) Linolinic
(C) Linoleic (D) All of these

431. Which is most common rooting hormone?
(A) ABA (B) IBA
(C) NAA (D) GA

432. Rose wintering is very common in which region?
(A) Calcutta (B) Pune
(C) Delhi (D) Shimla

433. Which is a growth inhibitor?
(A) MH (B) SADH
(C) B-Nine (D) All of these

434. Who is known as father of botany?
(A) Collet (B) Shull
(C) Theophrastus (D) Darwin

435. Second largest genera of Orchid?
(A) Bulbophyllum (B) Dendrobium
(C) Vanda (D) Oncidium

436. Which species of Orchid is grown on trees?
(A) Lithophyte (B) Epiphyte
(C) Saprophyte (D) Terrestrial

437. Pusa Krishna is a variety of ....
(A) Helichrysum (B) Hollyhock
(C) Statice (D) Stock

438. Which variety is not a Tuberose cultivar?
(A) Shringar (B) Suvasini
(C) Swarna (D) Rajat Rekha

439. Motia cultivar of Jasmine is related to ....
(A) *Auriculatum* (B) *Sambac*
(C) *Grandiflorum* (D) *Primulinum*

440. Manali is a variety of which flower?
(A) Hibiscus (B) Gladiolus
(C) Dahlia (D) Gerbera

441. Which of the following is a hardiest annual?
(A) Antirrhinum (B) Digitalis
(C) Stock (D) Aster

442. Which colour flower is found in Ageratum?
(A) White (B) Pink
(C) Blue (D) Red

443. Blitox is an ....
(A) Insecticide (B) Fungicide
(C) Weedicide (D) Bactericide

444. Which of the following is not grown for foliage purpose?
(A) Coleus (B) Kochia
(C) Ageratum (D) Celosia

445. Anemophily is the process of pollination through ....
(A) Insect (B) Water
(C) Wind (D) Birds

446. Which of the following is originated from India?
(A) Balsam (B) Lady's lace
(C) Gomphrena (D) All of these

447. *Michelia champaca* is originated from?
(A) America (B) Mexico
(C) India (D) South Africa

448. Which of the following is not a climber?
(A) *Ficus repens* (B) *Porana paniculata*
(C) *Hedera helix* (D) *Bougainvillea sp.*

449. How many flowering plants species reported in India?
(A) 10000 (B) 12000
(C) 15000 (D) 20000

450. Money maker is the cultivar of ....
(A) Lotus (B) Chrysanthemum
(C) Jasmine (D) Gerbera

451. Rocket is a F1 hybrid of which flower?
(A) Pansy (B) Petunia
(C) Antirrhinum (D) Coleus

452. Cinderella character is common in ....
(A) Aster (B) Cosmos
(C) Begonia (D) Petunia

453. Char Bagh is located at ....
(A) London (B) Iran
(C) Iraq (D) Turkey

454. How many doors are allowed in Baradari?
(A) 6 (B) 12
(C) 18 (D) 15

455. Who was started tissue culture in Orchid?
(A) Chang (B) Morel
(C) Lysenko (D) Shull

456. Which is not a common name of Lotus in Sanskrit?
(A) Kamala (B) Water lily
(C) Pankaja (D) Padma

457. Who is associated with Lotus
(A) Brahma (B) Shiva
(C) Krishna (D) Botha A & B

458. Carnation is an .... plant
(A) Long day (B) Short day
(C) Day neutral (D) Short – long day

459. *Rosa rugose* is resistant to ....
(A) Heat (B) Cold
(C) Both A & B (D) None of these

460. Fragrance is controlled by ....
(A) Mono gene (B) Poly gene
(C) Oligo gene (D) All of thcsc

461. .... cultivar of Rose is grown in greenhouse
(A) Cocktail (B) Jacaranda
(C) Konfetti (D) All of these

462. Pyrethrum is extracted from .... Chrysanthemum species
(A) *indicum* (B) *sinensis*
(C) *morifolium* (D) *cinerariifolium*

463. .... is used for disbudding in Chrysanthemum
(A) Cycocel (B) MH
(C) SADH (D) Oxathin

464. Basic chromosome number in Chrysanthemum is ....
(A) 8 (B) 9
(C) 12 (D) 10

465. Which of the following is responsible for phyllody disease in Jasmine?
(A) Mycoplasma (B) Virus
(C) Bacteria (D) Fungus

466. Who did introduced climate controlled green house in India?
(A) Sutton & Sons (B) ICAR
(C) Mahyco seed comp. (D) Indo-America hybrid seed co.

467. Nicky is a hybrid cultivar of ………….
(A) Nicotiana (B) Primula
(C) Petunia (D) Rose

468. In ………….. form Amaryllis species are self-fertile
(A) Triploid (B) Diploid
(C) Tetraploid (D) Haploid

469. Gerbera belong to ………….. family
(A) Liliaceae (B) Fabaceae
(C) Asteraceae (D) Ranunculaceae

470. Statice is a popular ………..
(A) Dry flower (B) Pot plant
(C) Bedding plant (D) All of these

471. Which is known as golden rod?
(A) Pansy (B) Solidago
(C) Stock (D) Amaranthus

472. Solidago is native from ………..
(A) South Africa (B) America
(C) India (D) Chin

473. Strelitzia is belong to which family?
(A) Cannaceae (B) Iridaceae
(C) Sterlitziaceae (D) None of these

474. *Callistephus chinensis* is an botanical name of ……….
(A) China soe flower (B) China Aster
(C) Gurhal (D) All of these

475. Stock behave in hilly areas like as ………..
(A) Perennial (B) Biennial
(C) Annual (D) Both B & C

476. Blanket flower is common name of ……….
(A) Calendula (B) Gaillardia
(C) Gomphrena (D) Limonium

477. *Narcissus pseudonarcissus* is botanical name of ……….
(A) Digitalis (B) Daffodil
(C) Primula (D) Cyclamen

478. Which of the following sterility commonly found in Ornamental plants?
(A) Cytoplasmic genetic (B) Genetic
(C) Cytoplasmic (D) All of these

479. Self – incompatibly is induced by ....
(A) Irridation (B) Surgical techniques
(C) Bud pollination (D) All of these

480. Maximum somatic hybrids have been released in ....
(A) Petunia (B) Stock
(C) Geranium (D) None of these

481. Which colour cultivars of Petunia evolve by genetic engineering?
(A) Yellow (B) Orange
(C) Blue (D) Purple

482. Who was planned Chandigarh city landscaping?
(A) Meyer (B) Maxwell
(C) Corbusier (D) M.S. Randhawa

483. Which tree is not suitable for avenue planting in plains?
(A) *Bassia latifolia* (B) *Eucalyptus hybirda*
(C) *Azadirachta indica* (D) *Dalbergia sissoo*

484. Important character of plants for avenue planting in plains?
(A) Brittle branches (B) Shade
(C) Thorny branches (D) Columnar crown

485. Which tree is most suitable for planting in Rajasthan?
(A) *Albizia lebbek* (B) *Butea monosperma*
(C) *Prosopis juliflora* (D) *Pongamia glabra*

486. Crocus bulb is stored at .... °C
(A) 1-2 (B) 10-12
(C) 13 (D) 4-5

487. First of all Breaking virus reported in ....
(A) Rose (B) Tulip
(C) Lily (D) Gladiolus

488. Begonia is native from ....
(A) South America (B) South America
(C) India (D) Peru

489. Which is belong to family Asteraceae?
(A) Tuberose (B) Dahlia
(C) Begonia (D) Sweet William

490. Mostly how to use lighting in garden?
(A) Walkway (B) Shadow
(C) Silhouette (D) None of these

491. How to create Rhythm in a garden?
(A) Progression of sizes (B) Continuous line movement
(C) Repetition of shapes (D) All of these

492. Which of the following has fine texture?
(A) Silver oak (B) Deodar
(C) Jacaranda (D) All of above

493. In garden Mobility is created by use of ....
(A) Evergreen trees (B) Evergreen shrubs
(C) Deciduous tree (D) Different size trees

494. How much area devoted for shrubbery borders in home garden?
(A) 15-20% (B) 5-10%
(C) 10-15% (D) 20-25%

495. Which is not a type of texture?
(A) Line (B) Coarse
(C) Fine (D) Medium

496. Garden style selection depends upon ....
(A) Space (B) Location
(C) Topography (D) All of these

497. Largest importer of house plant is ....
(A) Germany (B) Netherland
(C) Norway (D) Italy

498. Optimum p^H of potting media for ferns growing?
(A) 3-4 (B) 5-7
(C) 5-6 (D) 5.5-6.5

499. Lack of air in potting media is due to ....
(A) Over watering (B) Inadequate pore space
(C) Compacting (D) All of above

500. Cheapest container is ....
(A) Copper (B) Plastic
(C) Earthen (D) Concrete

501. Which of the following is not a deciduous bulbous plant?
(A) Corcus (B) Dahlia
(C) Watsonia (D) Tulip

502. Spectabilis is a variety of ....
(A) Sweet pea (B) Sweet Sultan
(C) Sweet Allysum (D) Sweet William

503. Optimum soil temperature of potting media?
(A) 10-20 (B) 18-21
(C) 15-20 (D) 20-25

504. Aralia is propagated through ....
(A) Cutting (B) Seed
(C) Corm (D) All of these

505. Cane cutting is commonly found in ....
(A) Coleus (B) Diefenbachia
(C) Philodendron (D) Araucaria

506. Air layering is not commonly found in ....
(A) Begonia (B) Croton
(C) Aglaonema (D) Dracaena

507. Mound layering is commonly used in ....
(A) Cestrum (B) Cactus
(C) Hydrangea (D) Agave

508. With high concentration of sugar treatment of cut flowers is known as ....
(A) Hardening (B) Pulsing
(C) Loading (D) Lodging

509. Optimum room temperature for pulsing treatment?
(A) 10-12 ºC (B) 15-20 ºC
(C) 20-25 ºC (D) 10-15 ºC

510. Cut flower impregnation is very common in ....
(A) Gerbera (B) Aster
(C) Carnation (D) All of these

511. Euphorbia is .... Sensitive to ethylene
(A) Moderately (B) Highly
(C) Not (D) Less

512. Rose is .... sensitive to chilling injury
(A) Less (B) Highly
(C) Moderately (D) Not

513. Cold storage temperature for carnation should be .... ºC
(A) 1-2 (B) 0-1
(C) 8-10 (D) 4-5

514. Which of the following is not suitable for dry transportation?
(A) Chrysanthemum (B) Iris
(C) Carnation (D) Lily

515. Which of the following is a bactericide?
(A) 8 HQC (B) DICA
(C) Both A & B (D) Citric acid

516. Bud blasting in Iris is due to ....
(A) Water stress (B) Poor light
(C) High temperature (D) All of these

517. St. Augustine grass has which type texture?
(A) Fine (B) Medium
(C) Coarse (D) Rough

518. Red fescue grass .... Shade tolerant
(A) Less (B) Medium
(C) Highly (D) Very less

519. Tall fescue is highly resistant to ....
(A) High RH (B) Drought
(C) Chilling (D) All of these

520. Creeping bent has .... texture
(A) Fine (B) Coarse
(C) Medium (D) None of these

521. *Ailanthus excels* is originated from............
(A) India (B) Japan
(C) North Korea (D) North America

522. In *Amherstia* flowering occurs during ....
(A) Feb-April (B) Feb-March
(C) Dec-March (D) Sept-Dec

523. *Chorisia* has value for beautiful............
(A) Flower (B) Fruit
(C) Trunk (D) Foliage

524. *Erythrina indica* is a botanical name of ....
(A) Coral tree (B) Coral vine
(C) Amaltas (D) Siris

525. *Polyalthia longifolia* is planted .... in multistory building
(A) Back (B) Back & front
(C) Around (D) Front

526. *Saraca indica* is associated with ....

(A) Sita (B) Kali

(C) Sarashwati (D) All of these

527. *Polyalthia longifolia* produce flowers during ....

(A) October (B) November

(C) March (D) July

528. *Dillenia indica* has which colour flower?

(A) Red (B) White

(C) Yellow (D) Orange

529. *Rhododendron arboretum* is suitable for ....

(A) Lower hills (B) Middle hills

(C) High hills (D) Plains

530. Palas (flame of forest) bear flowers on ....

(A) New branches (B) Old branches

(C) Leaf less branches (D) All of these

531. *Plumeria rubra* has .... colour flower

(A) White (B) Red

(C) Blue (D) Yellow

532. Bougainvillea is climbs through ....

(A) Hooks (B) Tendrils

(C) Petiole (D) None of these

533. Which species of *Tecoma* is not a climber?

(A) *Capensis* (B) *Grandiflora*

(C) *Stans* (D) All of these

534. *Trachelospermum jasminoides* is grow well in ....

(A) Dark (B) Shade

(C) Sunny situation (D) All of these

535. *Pyrostegia venusta* is belong to family ....

(A) Bignoniaceae (B) Comberateceae

(C) Fabaceae (D) None of these

536. *Hiptage benghalensis* is belong to which family?

(A) Moraceae (B) Malveceae

(C) Malpighiaceae (D) Hiptageaceae

537. *Smilax aspera* popular for beautiful ....

(A) Foliage (B) Fruits

(C) Flower (D) All of these

538. *Wisteria sinensis* is belong to which family?
(A) Comberateceae (B) Moraceae
(C) Fabaceae (D) Malveceae

539. Which is suitable for porches?
(A) *Ipomoea horsfalliae* (B) *Pyrostegia venusta*
(C) *Clerodendron splendens* (D) All of above

540. Which species of *solanum* is known as potato creeper?
(A) *tuberosum* (B) *jasminoides*
(C) *seaforthianum* (D) *wendlandii*

541. *Acer palmatum* has beautiful ....
(A) Flower (B) Fruit
(C) Foliage (D) Trunk

542. *Gynura aurantica* has .... Colour foliage
(A) Red (B) Green
(C) Violet (D) Purple

543. *Euphorbia cotinifolius* has which colour foliage?
(A) Coppery (B) Blue
(C) Orange (D) Red

544. *Mussaenda* has beautiful ....
(A) Flower (B) Foliage
(C) Bracts (D) Trunk

545. For rockeries purpose shrubs should be ....
(A) Drought tolerant (B) Moisture loving
(C) Deciduous (D) All of these

546. Popular pot plant in International market is ....
(A) Hibiscus (B) Ixora
(C) Crossandra (D) Poinsettia

547. *Breynia nivosa* is grown for ....
(A) Flower (B) Pot culture
(C) Cut greens (D) All of these

548. Acalypha hispida is belong to which family?
(A) Moraceae (B) Malveceae
(C) Euphorbiaceae (D) Asteraceae

549. *Russelia juncea* has .... colour flower
(A) White (B) Red
(C) Yellow (D) Blue

550. *Russelia juncea* has originated from ....
(A) South America (B) India
(C) Peru (D) North China

551. *Ochna jabotapita* has .... Colour fruit
(A) White (B) Black
(C) Green (D) Purple

552. Which of the following does not have yellow colour flower?
(A) *Bauhinia tomentosa* (B) *Pentas lanceolate*
(C) *Ochna jabotapita* (D) Both b & c

553. Which of the following type leaves found in *Virburnum cotinifolius*?
(A) Leathery (B) Shining
(C) Glaucous (D) Feather like

554. Din ka raja is a common name of ....
(A) *C. nocturnum* (B) *C. diurnum*
(C) *C. parqui* (D) None of these

555. *Raphis excelsa* is propagated through ....
(A) Air layering (B) Suckers
(C) Seed (D) All of these

556. Palms are mostly tolerant to ....
(A) High temperature (B) Frost
(C) High rains (D) None of these

557. *Roystonea regia* is a botanical name of ....
(A) Bottle palm (B) Royal palm
(C) Sago palm (D) Fish tail palm

558. How many genera are there in family Cycadaceae?
(A) 10 (B) 9
(C) 14 (D) 20

559. Which of the following is not belong to family Cycadaceae?
(A) Licuala (B) Zamia
(C) Cycas (D) Dioon

560. Another name of Selaginells is ....
(A) Club moss (B) Moss
(C) Lycopodium (D) All of these

561. Stag Horn fern is an another name of ....
(A) *Pellaea* (B) *Nephrolepis*
(C) *Platycerium* (D) *Polypodium*

562. How many species of ferns in the world?
(A) 1000 (B) 10000
(C) 15000 (D) 12000

563. .... is another name of *Echinocactus grusonii*
(A) Chin cactus (B) Golden Barrel
(C) Rainbow cactus (D) Tom's thumb

564. Bunny Ears is an another name of ....
(A) *microdosys* (B) *tetracantha*
(C) *mulgaris* (D) *tunicata*

565. Agave is also known as ....
(A) Money plant (B) Century plant
(C) Slipper plant (D) None of these

566. Slipper plant is an another name of ....
(A) *Greenovia* (B) *Pachyveria*
(C) *Pedilanthus* (D) *Paphiopedilum*

567. Cacti and succulents are classified as ....
(A) Lithophyte (B) Neophyte
(C) Xerophyte (D) Saprophyte

568. For growing of cacti water should be ....
(A) Neutral (B) Alkaline
(C) Acidic (D) Any of above

569. Rat tail cactus comes under genus ....
(A) *Ferocactus* (B) *Zyzocactus*
(C) *Aporocactus* (D) *Ariocactus*

570. What is means of xeric environment?
(A) Low rainfall (B) High temperature
(C) Low humidity (D) All of these

571. *Kalanchoe* is propagated through ....
(A) Stem cutting (B) Leaf cutting
(C) Slips (D) Suckers

572. *Agave* is propagated by ....
(A) Offsets (B) Cutting
(C) Seed (D) All of these

573. 'Indian fig' is another name of which species of Opuntia?
(A) *vilis* (B) *rufida*
(C) *falcate* (D) *ficus-indica*

574. 'Pretty Stone Face' is another name of ....
(A) Lithops (B) Senecio
(C) Sedum (D) Orbea

575. *Heliconia* belong to which family?
(A) Cannaceae (B) Bromeliadaceae
(C) Musaceae (D) Rutaceae

576. Suitable climate for Palm is ....
(A) Warm-humid (B) Warm-dry
(C) Cool-dry (D) Cool-humid

577. *Chamaedorea elegans* is a ....
(A) Palm (B) Cacti
(C) Fern (D) Ornamental tree

578. *Lycopodium* is a .... fern
(A) Spray type (B) Climbing type
(C) Both a & b (D) None of these

579. Which drying method is suitable for Pansy?
(A) Air drying (B) Press drying
(C) Vacuum drying (D) All of these

580. Which of the following is used for drying?
(A) River sand (B) Silica gel
(C) Borax (D) All of these

581. In Microwave flowers are dried for .... minutes
(A) 15-20 (B) 1-5
(C) 30-40 (D) 5-10

582. Where is the biggest market for dry flowers?
(A) U.K (B) Germany
(C) Netherland (D) USA

583. Maximum damage in Gladiolus is caused by ....
(A) Mites (B) Thrips
(C) Aphids (D) White fly

584. For Chrysanthemum which insecticide is not good?
(A) Rogor (B) Endosulphon
(C) Malathion (D) Metasystox

585. In greenhouse which insect is most serious?
(A) Aphid (B) White fly
(C) Larvae (D) Thrips

586. Orius bug is controls ....
(A) Thrips (B) White fly
(C) Aphid (D) All of above

587. Meghdoot is a variety of which flower?
(A) Marigold (B) Chrysanthemum
(C) Rose (D) Carnation

588. In Chrysanthemum cultivar Maghi flowering during ....
(A) Dec-Jan (B) Feb-March
(C) Oct-Nov (D) Nov-Dec

589. Which of the following is used for checked pollen sterility in Rose?
(A) Ethyl alcohol (B) Potassium Iodide
(C) Acetocarmine (D) All of these

590. Which of following is used for mutation studies?
(A) MES (B) NMU
(C) EMS (D) All of above

591. Gene mutation is commonly found in ....
(A) Phlox (B) Sweet pea
(C) China aster (D) All of these

592. Which flower is not used for garlands making?
(A) Chrysanthemum (B) Gladiolus
(C) Marigold (D) Crossandra

593. For making floral bangles which flower is used?
(A) Tuberose (B) Chandni
(C) Jasmine (D) All of these

594. .... is not used for making gajra
(A) Barleria (B) Marigold
(C) Michelia (D) Crossandra

595. Flower arranger world association was established in ....
(A) 1965 (B) 1982
(C) 1981 (D) 1990

596. At IIHR Bangalore Floriculture division was started during ....
(A) 1960 (B) 1969
(C) 1966 (D) 1970

597. At IARI New Delhi Floriculture division was established in ....
(A) 1974 (B) 1973
(C) 1983 (D) 1986

598. *Nymohaea* research has been conducted at ....
(A) NBRI (B) IIHR
(C) BSI Calcutta (D) IARI

599. Dr. Abraham was associated with which flower?
(A) Orchid (B) Carnation
(C) Pansy (D) Calendula

600. S.P.S. Raghava is associated with ....
(A) Dahlia (B) Marigold
(C) Jasmine (D) Chrysanthemum

601. Central Institute for Medicinal and aromatic Plants is located at ....
(A) Lucknow (B) Delhi
(C) Ajmer (D) Ahmedabad

602. NRC on Medicinal and Aromatic Plants is located at ....
(A) Jodhpur (B) Bhopal
(C) Anand (D) Kolkata

603. NRC on Orchid is located at ....
(A) Sikkim (B) Pondicherry
(C) Dehradun (D) Himachal

604. Bonsai is a .... term
(A) Chinese (B) French
(C) Japanese (D) Arabic

605. Bonsai technique was developed .... Years ago
(A) 2000 (B) 1500
(C) 1000 (D) 800

606. What is optimum size for miniature Bonsai?
(A) >15 cm (B) <10 cm
(C) <15 cm (D) $\leq$ 8 cm

607. A Bonsai with single straight stem and lateral alternative branches on both side is Known as ....
(A) Formal upright (B) Informal upright
(C) Winding (D) Cascade

608. Which ratio of garden soil and leaf mould is suitable for Bonsai culture?
(A) 2: 1 (B) 3 : 1
(C) 2: 3 (D) 3 : 4

609. The failure of a bulb to produce any floral parts is known as ....
(A) Blind shoot (B) Blindness
(C) Bent neck (D) All of these

610. Modified leaf, frequent associated with flower is called as ....

(A) Axil (B) Axis

(C) Bracts (D) Basipetal

611. Spherically shaped flower bud, which is produce by malformation of flower

(A) Bullhead (B) Bullnose

(C) Bulkers (D) Blind shoot

612. The process of yellowing of foliage due to loss of chlorophyll

(A) Damping off (B) Chlorosis

(C) Chimera (D) Conidia

613. The trumpet or cup of a Daffodil flower is called as ....

(A) Corolla (B) Crown

(C) Corona (D) Coppice

614. Death of shoots, originating at the shoot tip is known as ....

(A) Blind shoot (B) Die back

(C) Wilting (D) Chlorosis

615. Removal of lateral flower buds on stem of plant is known as ....

(A) Disshooting (B) Disbudding

(C) Denovelling (D) All of these

616. Fleck is a common disease of ....

(A) Gladiolus (B) Tuberose

(C) Lily (D) All of these

617. A group of similar species is known as ....

(A) Genus (B) Genera

(C) Class (D) Family

618. Removal of the terminal shoot to promote side branching is called as ....

(A) Pruning (B) Pinching

(C) Disshooting (D) Pinking

619. Shattering is the process of abscission of .... Flower

(A) Lily (B) Pansy

(C) Snapdragon (E) Gladiolus

620. A mutant that is inherited and transmitted to progeny is known as ....

(A) Sport (B) Sporulate

(C) Spores (D) Sours

621. Which of the following was introduced by British in India?

(A) Aster (B) Carnation

(C) Larkspur (D) All of above

622. Buddha Jayanti Park New Delhi was established in ....
(A) 1950 (B) 1956
(C) 1959 (D) 1965

623. Rose Garden in Chandigarh was laid out in ....
(A) 1963 (B) 1960
(C) 1966 (D) 1970

624. Removal of the above ground growth of grass is called as ....
(A) Rolling (B) Mowing
(C) Scrapping (D) Breaking

625. Siris is originated from ....
(A) Tropical Asia (B) Tropical America
(C) Tropical Africa (D) Tropical Australia

626. Tree of Haven is also known as ....
(A) Monkey bread (B) Paradise tree
(C) Cook pine (D) Hoop pine

627. Golden Shower is belong to which family?
(A) Fabaceae (B) Apocynaceae
(C) Araucaria (D) Rubiaceae

628. *Artocarpus altilis* is botanical name of ....
(A) Margosa tree (B) Breadfruit tree
(C) Kachnar (D) Indian oak

629. Bottle brush belong to family ....
(A) Myrataceae (B) Fabaceae
(C) Malvaceae (D) Bombacaceae

630. Banyan tree is originated from ....
(A) America (B) South Africa
(C) India (D) Maxico

631. Piple tree is belong to family?
(A) Moraceae (B) Rubiaceae
(C) Rutaceae (D) Malvaceae

632. Royal Palm is originated from ....
(A) Peru (B) Cuba
(C) Brazil (D) Malaysia

633. Which of the following have drooping habit?
(A) *Lantana depressa* (B) *Plumbago capensis*
(C) *Russelia juncea* (D) All of these

634. Which of the following don't have bushy habit?
(A) *Barleria cristata* (B) *J. sambac*
(C) *Hamelia patens* (D) *Malpighia coccigera*

635. Which of the following is used for Topiary?
(A) Duranta (B) Clerodendrum
(C) Hibiscus (D) All of above

636. Which is not used for Ornamental foliage?
(A) Dracaena (B) Juniperus
(C) Thuja (D) Poinsettia

637. Which of the following does not have fragrant flower?
(A) Pedilanthus (B) Gardenia
(C) Magnolia (D) Buddlea

638. *Thunbergia exotica* ………… Colour flowering shrub
(A) White (B) Blue
(C) Yellow (D) Red

639. *Cassia alata* is an ………….. colour flowering shrub
(A) Yellow (B) Pink
(C) Red (D) White

640. *Mussaenda spp*. Shrub is popular for …………
(A) Fragrant flower (B) Bracts
(C) Pot plants (D) Both A & B

641. Acalypha is belong to family ………..
(A) Amaranthaceae (B) Euphorbiaceae
(C) Scrophulariaceae (D) Acanthaceae

642. Rosy Carpet is a cultivar of …………
(A) Lantana (B) Lagerstroemia
(C) Jatropha (D) Juniperus

643. Tithonia is also known as ………..
(A) Mexican Sunflower (B) Tree Marigold
(C) Both A & B (D) Chinese rose

644. *Monstera deliciosa* is a ………….. type climber
(A) Flowering (B) Foliage
(C) Screening (D) All of above

645. *Antogonon leptopus* is a ……….. colour flowering climber
(A) Red (B) Pink
(C) White (D) Yellow

646. *Bigonia venusta* have .... Colour flowering climber

(A) Yellow (B) Pink

(C) Blue (D) Orange

647. Black Eye susan is a common name of ....

(A) *T. alata* (B) *T. fragrans*

(C) *T. grandiflora* (D) *T. mysorensis*

648. Which of the following is not a summer annual?

(A) Gaillardia (B) Kochia

(C) Balsam (D) Cosmos

649. Which of the following is a winter annual?

(A) Phlox (B) Marigold

(C) Tithonia (D) Zinnia

650. Candytuft is a .... colour flowering annual

(A) Blue (B) Pink

(C) White (D) Yellow

651. Which colour flowers are found in Coreopsis?

(A) Blue (B) Pink

(C) White (D) Yellow

652. Which of the following is not a late blooming annual?

(A) Antirrhinum (B) Phlox

(C) Pansy (D) Petunia

653. Which of the following is an early blooming annual?

(A) Carnation (B) Hollyhock

(C) Balsam (D) Sweet William

654. Which is not have Ornamental foliage?

(A) Sweet Alyssum (B) Amaranth

(C) Kochia (D) Nasturtium

655. Which of the following is not used for rockery purpose?

(A) Brachycome (B) Portulaca

(C) Sweet Pea (D) Verbena

656. Pansy is not useful for .... Purpose

(A) Hanging Basket (B) Carpet Bedding

(C) Edging (D) Rockery

657. Which of the following is an fragrant flowering annual?

(A) Sweet Pea (B) Sweet William

(C) Sweet Sultan (D) All of these

658. Floss flower is an .... colour flowering annual

(A) Red (B) White

(C) Blue (D) Pink

659. Hollyhock belong to family ....

(A) Asteraceae (B) Malvaceae

(C) Cruciferae (D) Boraginaceae

660. *Ceantaurea cyanua* is botanical name of............

(A) Corn flower (B) Cone flower

(C) Sweet Sultan (D) Sweet Potato

661. Which of the following is not belong to family Asteraceae?

(A) Corn flower (B) Sweet sultan

(C) Wall flower (D) Gazania

662. *Acalypha hispida* required .... light intensity

(A) High (B) Medium

(C) Low (D) Any of above

663. House plant which is requiring low light intensity?

(A) Alternanthera (B) Aglaonema

(C) Rubber plant (D) Kalanchoe

664. *Chrysanthemum cinerariifolium* and *C. coccineum* are also sources of which important insecticide?

(A) Pyrethrum, an insecticide (B) Lutein, a xanthophyll

(C) delphinidin (D) none

665. *Chrysanthemum morifolium* is commonly known as.............

(A) *Florists' chrysanthemum* (B) Chinese/ Japanese mums

(C) Garland chrysanthemum (D) Tricolour chrysanthemum.

666. Chrysanthemum is a .... with respect to flowering.

(A) Qualitative Short day plant (B) Long day plant

(C) Short long day plant (D) Day neutral plant

667. Pigment responsible for blue colour in rose is ....

(A) Pelargonidin (B) Cynidine

(C) flavonols (D) delphinidin

668. Pruning of rose is done in the month ....

(A) Oct-Nov (B) Dec-Jan

(C) Feb- Mar (D) April-May.

669. Limp neck/Bent neck is a physiological disorder of which flower?

(A) Chrysanthemum (B) Tuberose

(C) Rose (D) Marigold

670. Which of the following is commonly known as sword lily?
(A) Gladiolus (B) Chrysanthemum
(C) Gerbera (D) Marigold

671. 100g rose syrup will contain ____________mg ascorbic acid
(A) 150 (B) 100
(C) 200 (D) 50

672. Most popular variety of rose in India for protected cultivated is ....
(A) Grand gala (B) Konfetti
(C) First Red (D) All of these

673. On the occasion of IARI 2016 convocation which variety of African marigold is released By President Pranab Mukherjee?
(A) Pusa Basanti gainda (B) Pusa Narangi Gainda
(C) Pusa Arpita (D) Pusa Bahar

674. Generic name of all Dahlias is ....
(A) *Dahlia variabilis* (B) *Dahlia rosea*
(C) *Dahlia juarezii* (D) *Dahlia imperialis*

675. Which of the following is national flower of India?
(A) *Nelumbo nucifera* (B) *Nelumbo aureavallis*
(C) *Nelumbo lutea* (D) None of the

676. The major carotenoid pigments present in marigold are ....
(A) Lutein (B) caricaxanthin
(C) lycopene (D) carotene

677. Which flower is known as "Queen of flowers" ....
(A) Chrysanthemum (B) Rose
(C) Dahlia (D) Gerbera

678. Dahlia has classified in how many groups by National Dahlia Society of England?
(A) 10 (B) 11
(C) 5 (D) 12

679. "Sen Rin Tsukuri" means growing .... chrysanthemum blooms together.
(A) 500 (B) 1000
(C) 1500 (D) 1200

680. "Kiku" is the Japanese art of training ....
(A) Chrysanthemum (B) Rose
(C) Dahlia (D) Gerbera

681. Progenitor/ancestor of yellow coloured roses is ....
(A) *Rosa foetida* (B) *Rosa canina*
(C) *Rosa gallica* (D) *Rosa hybrida*

682. First tissue culture in Orchid (Meristem culture) was discovered by ....
(A) George Morel (B) Noel Bernard
(C) George Mendel (D) Hans Burgeff

683. At the time of departure which of the following is presented?
(A) Marigold (B) Lilium
(C) Sweet pea (D) Tulip

684. Most suitable anthurium species for pot cultivation is ....
(A) *A. andreanum* (B) *A. scherzerianum*
(C) *A. hookeri* (D) *A. crystallinum*

685. Major disease of Anthurium is ....
(A) Anthracnose (B) Leaf spot
(C) Root rot (D) Bacterial blight

686. Foliage capital of the world is ....
(A) Alsmeer (B) Apopka
(C) Las Vegas (D) Chicago

687. Largest producer and exporter of anthurium is ....
(A) USA (B) Netherlands
(C) Mauritius (D) Australia

688. Second largest producer and exporter of anthurium is ....
(A) USA (B) Netherlands
(C) Mauritius (D) Australia

689. Which species of Anthurium is known as flame plant or flamingo flower?
(A) *A. andreanum* (B) *A. scherzerianum*
(C) *A. hookeri* (D) *A. crystallinum*

690. Required light intensity for Anthurium?
(A) 1000-1500 Lux (B) 20000-25000 Lux
(C) 25000-30000 Lux (D) 30000-40000 Lux

691. Flower of Anthurium is ....
(A) Androgynous (B) Hypogynous
(C) Protogynous (D) Epigynous

692. Optimum light intensity required for Orchid ....
(A) 3000-6000 foot candle (B) 2000-4000 foot candle
(C) 4000-6000 foot candle (D) 5000-7000 foot candle

693. Upright bonsai style is also known as ....

(A) Kengai (B) Han-Kengai

(C) Chokkan (D) Shakan

694. Lawn is the main feature of which type garden?

(A) Japanese (B) Mughal

(C) English (D) Persian

695. Who was discovered wild garden style?

(A) William Robinsan (B) William Gaud

(C) Repton (D) Le Notre

696. What is the mean of Japanese word Tsukiyama-niwa?

(A) Hill and Water (B) Hill and Tree

(C) Water and Tree (D) Stones and Tree

697. Which style garden designed on the base of heaven?

(A) Mughal (B) Japanese

(C) Persian (D) Both b & c

698. Which of the following is a feature of Japanese garden?

(A) Waterfalls (B) Wells

(C) Islands (D) All of these

699. Pagodas is the main feature of .... garden

(A) Mughal (B) Japanese

(C) Persian (D) Italian

700. Vertical garden is developed from ....

(A) Australia (B) Switzerland

(C) China (D) USA

701. Lotus is a .... type of plant

(A) Surface flowering (B) Oxygenators

(C) Marginals (D) Floaters

702. Which of the following plant is an oxygenator?

(A) Azolla (B) Lotus

(C) Elodea (D) All of these

703. Water butter-cup is a .... Type plant

(A) Surface flowering (B) Oxygenators

(C) Marginals (D) Floaters

704. Which is the principle of the relative dimension?

(A) Balance (B) Scale

(C) Space (D) Harmony

705. In India, the theme of bio-aesthetic planning was propagated by ....
(A) M. S. Randhawa (B) Prof. Lancelot Hogben
(C) E. Maxwell (D) Jane B. Drew

706. CAD was developed by ....
(A) Dr. Patrick J. Hanratty (B) Jane B. Drew
(C) E. Maxwell (D) Repton

707. Carnation is also known as ....
(A) Divine flower (B) Gilly flower
(C) Clove flower (D) All of these

708. Candy is the variety of which flower?
(A) Chrysanthemum (B) 1Carnation
(C) Dahlia (D) Gerbera

709. Roma is a .... colour variety of carnation
(A) Pink (B) White
(C) Red (E) Yellow

710. Cuttings normally develop good root system within
(A) 10 days (B) 21 days
(C) 15 days (D) 30 days

711. Concentration of IBA used for rooting in cutting of carnation is
(A) 100 ppm (B) 400 ppm
(C) 500 ppm (D) 700 ppm

712. Soil P^H will be .... by the addition of sulphur
(A) Neutralizes (B) Increases
(C) Decreases (D) none of these

713. Which of the following carnation variety is the standard type?
(A) Excel (B) Candy white
(C) Rony (D) Silvery Pink

714. Which variety is developed through cross between Shringar X Maxican single?
(A) Vaibhav (B) Prajwal
(C) Suvasini (D) Rajat Rekha

715. Fresh flowers of tuberose give a concrete yield of
(A) 0.04-0.5% (B) 0.25-0.5%
(C) 0.08 to 0.11 % (D) 0.5-0.75%

716. Seeds of Tuberose are .... shaped
(A) Round (B) Oblong
(C) Flat (D) None of these

717. Tuberose oil contains have ....
(A) Methyl benzoate (B) Methyl anthranilate
(C) Benzyl alcohol (D) All of these

718. Crossandra belong to which family?
(A) Moraceae (B) Malveceae
(C) Acanthaceae (D) Asteraceae

719. Orange crossandra is an ....
(A) Triploid (B) Tetraploid
(C) Diploid (D) Aneuploidy

720. Chromosome number in Delhi crossandra are 2n = ...
(A) 20 (B) 30
(C) 40 (D) 10

721. Flowering in crossandra comes .... Months after planting
(A) 2-3 (B) 2.5-3
(C) 4-5 (D) 1-2

722. Average yield of Delhi crossandra is .... Kg of flowers
(A) 2000 (B) 2500
(C) 2800 (D) 3500

723. Which of the following plant is suitable for shade garden?
(A) Fox glove (B) Arrowhead
(C) Bull Rush (D) All of above

724. Duckweed is an .... type plant
(A) Floating (B) oxygenators
(C) Deep water (D) Submerged

725. Adiantum capillus is the botanical name of ....
(A) Venus hair fern (B) Bird nest fern
(C) Table fern (D) Tree fern

726. Which of the following is cycads?
(A) Adiantum (B) Caryota urens
(C) Pritchardia (D) Zamia

727. The word "ROSE" is derived from .... Word
(A) Latin (B) Greek
(C) French (D) None

728. Which type roses are suitable for hedges?
(A) miniatures (B) Floribundas
(C) Button roses (D) All of these

729. Dried rose petals are known as ....
   (A) Pankhuri (B) Gulroghan
   (C) potpourri (D) Both a & c
730. The chief aim of landscaping industrial areas is to reduce ....
   (A) wind velocity (B) noise
   (C) pollution (D) All of above
731. Which of the characters' desirable for industrial area plants?
   (A) Pubescence (B) Large number of stomata
   (C) Broad leaves with rough surface (D) All of these
732. Which of the following tree is SO2 tolerant?
   (A) Casuarina (B) Albizzia
   (C) Delonix regia (D) All of these
733. Flowers produce in gerbera .... weeks after planting
   (A) 8-10 (B) 7-8
   (C) 10-12 (D) 5-6
734. Major cut flower growing states is ....
   (A) Tamil Nadu (B) West Bangal
   (C) Karnataka (D) Punjab
735. Which of the following multi bracted cultivar of bougainvillea is
   (A) Lady Mary Baring (B) Golden Glow
   (C) Cherry Blossom (D) All of these
736. Bougainvillea cultivar 'Mary Palmer was developed from
   (A) Hybridization (B) Mutation
   (C) Introduction (D) None of above
737. Polyploidy is induced using
   (A) Colchicine (B) Maleic Hydrazide
   (C) GA (D) NAA
738. Dwarfing in ornamentals is induced by growth retardant like
   (A) B-Nine (B) Phosphone
   (C) Cycocel (D) All of these
739. The concept of bio-aesthetic planning in India was given by
   (A) M. S. Randhawa (B) Lancelot Hogben
   (C) Julius (D) All of these
740. Tree __________ is closely associated with the life of Lord Krishna.
   (A) Kadamba (B) Pipal
   (C) Ashok (D) Sal

741. The birth of Lord Buddha took place under the tree ....
(A) Asoka (B) Pipal
(C) Sal (D) Palas

742. Nandanavana garden is dedicated to Lord ....
(A) Shiva (B) Buddha
(C) Krishna (D) Kali

743. The Buddha attained his enlightenment under
(A) Pipal tree (B) Kadamba tree
(C) Asoka tree (D) None of above

744. Who has established the Red Fort in Delhi and Taj Mahal in Agra.
(A) Akbar (B) Jahangir
(C) Shahjahan (D) Sher Shah Suri

745. The Harsh Charita was written by
(A) Bana Bhatta (B) Vatsayana
(C) Asvaghosha (D) None of above

746. The first evidence of which ornamental tree comes from a seal from Mohan-Jo- Daro.
(A) Banyan tree (B) Asoka tree
(C) Arjun tree (D) Pipal tree

747. The type of Japanese garden which is completely devoid of plants is ....
(A) Zen garden (B) Tea garden
(C) Hill garden (D) Flat garden

748. Flowering annuals is a common feature of .... style gardens.
(A) English (B) Japanese
(C) Persian (D) Mughal

749. How many varieties of bougainvillea have been developed in India
(A) 100 (B) 150
(C) 75 (D) 125

750. French garden art was developed by ....
(A) Lepton (B) Capability
(C) Le Notre (D) Hogbean

751. Samrat a tetraploid cultivar of ....
(A) Rose (B) Amaryllis
(C) Verbena (D) Tuberose

752. Which one of the following is associated with Lord Buddha?
(A) Asoka tree (B) Jasmine
(C) Silk Cotton tree (D) Narcissus

753. Which one of the following is a formal style of garden?
(A) Hindu-Buddhist Garden (B) Japanese Garden
(C) Mughal Garden (D) English Garden

754. The tree associated with Lord Shiva is ....
(A) Jasmine tree (B) Kadamba tree
(C) Silk cotton tree (D) None of above.

755. Visual equilibrium of any landscape design is known as ....
(A) Scale (B) Unity
(C) Balance (D) Harmony

756. .... is created in any garden to avoid monotonous view.
(A) Accent (B) Emphasis
(C) Proportion (D) Both A & B

757. The overall effect of various features is known as ....
(A) Harmony (B) Trophy
(C) Scale (D) Unity

758. The total mass of any object or plant in landscape design is termed as ....
(A) Balance (B) Proportion
(C) Unity (D) Line

759. Maximum somatic hybrids have been developed in ....
(A) Petunia (B) Geranium
(C) Stock (D) None of above

760. A three dimensional confined view of terminal building is known as
(A) Axis (B) Vista
(C) Circulation (D) None of above

761. Which of the following is an element of landscape gardening
(A) Pattern (B) Form
(C) Colour (D) All of above.

762. Which of the following is a neutral colour in colour scheme
(A) White (B) Grey
(C) Red (D) Both a & b

763. Tertiary colour are formed by mixing of
(A) Red and white (B) Green and orange
(C) Red and Blue (D) None of above.

764. Which of the following annuals are used as cut flowers?
(A) Carnation (B) Corn flower
(C) Sweet William (D) All of these

765. Which of the following annuals are used for making dry flower products?
(A) Statice (B) Helichrysum
(C) Nigella (D) All of above

766. According to photoperiodic requirement Amaranths is a ....
(A) Long day plant (B) Short day plant
(C) Day neutral (D) Long short day

767. Bougainvillea climbs with the help of special structure called ....
(A) Tendrils (B) Thorns
(C) Roots (D) Sticky substances

768. An example of rambler is ....
(A) Lonicera (B) *Ficus repens*
(C) *Quisqualis indica* (D) Bigonia

769. Which of the following climber has fragrant flowers?
(A) Jasmine (B) Honey suckle
(C) Both a & b (D) None of above

770. Which of the following climber can grow under shade?
(A) *Clerodendron splendens* (B) *Thunbergia grandiflora*
(C) *Jacquemontia pentantha* (D) All of above

771. Which of the following is a climber?
(A) *Ficus elastic* (B) *Ficus infectoria*
(C) *Ficus repens* (D) *Ficus religiosa*

772. Gomphrena is a
(A) Long day plant (B) Short long day plant
(C) Intermediate plant (D) Day neutral plant

773. Which of the following annual is used for screening purpose?
(A) Hollyhock (B) Sweet Pea
(C) Moon flower (D) All of above.

774. Marigold is propagated through ....
(A) Cutting (B) Seed
(C) Tissue culture (D) All of above

775. Rose is commercially propagated through ....
(A) Cutting (B) T budding
(C) Seed (D) Suckers

776. Patch budding is a propagation method of .... tree.
(A) *Abelia xgrandiflora* (B) Pecan
(C) Siberian peashrub (D) All of above

777. Dahlia is propagating by .... cuttings.
(A) Stem cuttings (B) Tuberous cutting
(C) Root cutting (D) Leaf cutting

778. .... is propagating by mound or stool layering.
(A) Rose (B) Jasmine
(C) Marigold (D) Gerbera

779. Which of the following flower crops are propagated through seed
(A) Chrysanthemum (B) Rose
(C) Tuberose (D) None of above

780. Rhododendron is propagated through
(A) Saddle grafting (B) Inarch grafting
(C) T-budding (D) All of the above

781. Flat grafting is most commonly used for propagation of ....
(A) Bulbous plants (B) Trees
(C) Cacti and succulents (D) Both a & c

782. Which of the following is a summer annual?
(A) Salvia (B) Sweet Pea
(C) Kochia (D) Sweet alyssum

783. *Magnolia grandiflora* is propagated through
(A) Seeds (B) Cutting
(C) Air layering (D) All of above.

784. F1 hybrid seed of petunia was first produced by .... Pvt.Ltd.
(A) Indian hybrid seed company (B) Indo-American hybrid seed company
(C) Maharashtra seed company (D) None of these

785. Which is an examples of self-pollinated seasonal flower?
(A) China Aster (B) Alstroemeria
(C) Tuberose (D) Heliconia

786. The pH of soil for seed production should be ....
(A) 5.5-7.0 (B) 5-7
(C) 6-7 (D) 4-5

787. Which of the following is used to checked pollen sterility in rose?
(A) Ethyl alcohol (B) Potassium iodide
(C) Acetocarmine (D) All of these

788. Which of the following has genetic male sterility?
(A) Marigold (B) Zinnia
(C) Ageratum (D) All of above

789. Which of the following has self-incompatibility

(A) Petunia (B) Pansy

(C) Both a and b (D) None of above

790. In which colour Pelargenidin anthocyanidin is present?

(A) Orange-red (B) Purple

(C) Green (D) Blue

791. Which gene action can control the inheritance of pigments?

(A) Epastasis (B) Additive

(C) Dominance (D) All of above

792. Which of the following is an often cross pollinated flower crop

(A) Antirrhinum (B) Aster

(C) Dahlia (D) All of above.

793. Which elements are used for building up the plant architecture?

(A) CHO (B) NPK

(C) Nitrogen (D) Ca, Mg, S

794. The criteria of essentiality for nutrient was proposed by ....

(A) Lysenko (B) Robert Browen

(C) Arnon and Stout (D) M.S. Swaminathan

795. How much % nitrogen is found in whole plant.

(A) 2 % (B) 3 %

(C) 4 % (D) 6 %

796. The interveinal chlorosis in carnation is a symptom of ....

(A) 'Fe' deficiency (B) 'B' deficiency

(C) 'Mn' deficiency (D) All of above

797. Sebaculis red is variety of ....

(A) Tuberose (B) Marigold

(C) Jasmine (D) Crossandra

798. Pale foliage in rose is a deficiency symptom of ....

(A) Fe (B) Mn

(C) B (D) Cu

799. Phosphorous is available in which of the following form?

(A) H_2PO_4 (B) HPO_4^{2-}

(C) Both a & b (D) None of above

800. The total quantity of boron in plants is found as

(A) 25-75 ppm (B) 10-1500 ppm

(C) 3-150 ppm (D) 5-1500 ppm.

801. RAPD stands for ....
(A) Random Accumulation of Polymorphic DNA'
(B) Random Amplification of Polymorphic DNA'
(C) Random Amplification of Polyethylene DNA'
(D) Random Atmosphere of Polymorphic DNA'

802. Black spot of Rose is caused by ....
(A) Fungi (B) Bacteria
(C) Nutrient deficiency (D) All of above

803. The gene responsible for crimson coloured pelargonidin pigment in petunia was transferred from ....
(A) Barley (B) Zea Mays
(C) Marigold (D) Pearl millet

804. Bt transgenic are highly effective against .... pests of several crops.
(A) Caterpillars (B) Locust
(C) Grass hopper (D) All of these

805. Plant tissue culture is based on the principle of
(A) Explant (B) Organ
(C) Totipotency (D) None of above

806. Chi trans-gene in carnation provide resistance against
(A) Calyx splitting (B) Viruses
(C) Fusarium wilt (D) Aphids.

807. Floral binding protection gene is incorporated to modify the flower shape and colour in
(A) Gerbera (B) Carnation
(C) Marigold (D) Tuberose

808. Which of the following marker not a DNA based
(A) RAPD (B) STS
(C) Isozyme (D) CAPs

809. Which of the following method is direct gene transfer method?
(a) *Agrobacterium* based method (b) Particle gun method
(c) Using liposome (d) Both b & c.

810. 'Grand Gala' is a variety of ____________ and is suitable for greenhouse cultivation.
(A) Gerbera (B) Rose
(C) Carnation (D) Gladiolus

811. Which of the following are temporary structure?
(A) Cold frames (B) Shade nets
(C) Poly house (D) All of above

812. Mite is a most serious pest of
(A) Lilium (B) Carnation
(C) Gerbera (D) None of above

813. Planting time of Lilium in plains is ……………
(A) Feb-March (B) May-June
(C) Oct-Nov (D) All of above

814. 'Connecticut King' is a variety of
(A) Rose (B) Carnation
(C) Asiatic lily (D) Oriental lily.

815. Which of the following greenhouse has rigid cladding material?
(A) Multi Span (B) Saw tooth Multi Span
(C) Both a & b (D) None of above.

816. Suitable plants for bottle garden ……..
(A) Club moss (B) Bougainvillea
(C) Petunia (D) Chlerodendeon

817. Which of the following plants are used for terrarium?
(A) Philodendron (B) Spider plant
(C) Parlor palm (D) All of above

818. Which annual is suitable for rock garden?
(A) Nasturtium (B) phlox
(C) Verbena (D) All of above

819. The garden where plants grown in a large bottle as in a terrarium is known as …….
(A) Bottle garden (B) Window garden
(C) Roof garden (D) Bog garden

820. Which of the flowering annual is suitable for roof gardening?
(A) Antirrhinum (B) Pansy
(C) Dahlia (D) All of above

821. The plant suitable for xeriscaping is
(A) *Calliandra eriophylla* (B) *Tagetes lemmonii*
(C) Both a and b (D) None of above

822. Which of the following is suitable for terrace gardening
(A) Portulaca (B) *Lantana sp.*
(C) *Thymus sp* (D) All of above

823. Xeriscaping is based upon the following principle
(A) Soil analysis and improvement (B) Plant selection
(C) Efficient irrigation (D) All of above

824. Interiorscaping improves
(A) Indoor air quality (B) Work performance
(C) Stress levels (D) All of above.

825. Leaf scorch injury in gladiolus is observed due to the presence of ....
(A) Floride (B) Zinc
(C) Potassium (D) Boron

826. The main constituents of floral preservatives are ....
(A) Sucrose (B) CaCO3
(C) NaCl (D) HCL

827. In CA, ____________% CO2 and _______% O2 is kept to get good quality flowers.
(A) 4%, 0.4% (B) 5%, 0.2%
(C) 6%, 0.4 (D) 3%, 0.1%

828. Short duration pre storage treatment with solution is known as ....
(A) Holding (B) Pulsing
(C) Pre-cooling (D) All of above

829. Microbial growth in vase water is decreases at pH
(A) 3.0-3.5 (B) 8-8.5
(C) 7.0 (D) None of above

830. Which of the following is act as biocide
(A) Sugar (B) Citric acid
(C) 8-HQC (D) Both b and C

831. The relative humidity inside the cold room is kept around
(A) 40-50 % (B) 60-70 %
(C) 90-95 % (D) None of above

832. Dry storage is also known as
(A) Modified atmosphere storage (B) Controlled atmosphere storage
(C) Low pressure storage (D) All of above.

833. Artificial colouring of the flowers to create novelty is called as __________
(A) Colouring (B) Tinting
(C) Pulsing (D) Dehydrating

834. Preserving the rose petals in sugar is known as __________
(A) Candy (B) Gulkand
(C) Surbat (D) Rose oil

835. Out of total floriculture exports, dry flower constitutes ________ share.
(A) 50% (B) 30%
(C) 60% (D) 80%

836. India is the ________ largest producer of essential oil
(A) 1st (B) 2nd
(C) 3rd (D) None of above

837. Which of the following country is the largest exporter of essential oil?
(A) India (B) USA
(C) Australia (D) Turkey

838. Gulkand is made from ....
(A) Rose (B) Chrysanthemum
(C) Periwinkle (D) None of above.

839. The origin of the following annual is not India
(A) *Gomphrena globose* (B) *Impatiens balsamina*
(C) *Salvia splendens* (D) *Vinca minor*

840. Which of the following is not valued for ornamental bracts?
(A) Bougainvillea (B) Poinsettia
(C) Mussaenda (D) Cyclamen

841. Work on Hibiscus improvement and development of varieties was extensively carried at
(A) IARI, New Delhi (B) NBRI, Lucknow
(C) IIHR, Bengaluru (D) YSRUHF, Solan

842. 'Star Gazer' is a poplar variety of
(A) Tiger Lily (B) Asiatic Lily
(C) Oriental Lily (D) Torch Lily

843. Which of the following is propagated through bulbils?
(A) Tulip (B) Lilium
(C) Gladiolus (D) Crocosmia

844. Botanically Kentucky blue grass is
(A) *Paspalum notatum* (B) *Agrostis palustris*
(C) *Poa pratensis* (D) *Festuca arundinacea*

845. Lawn mower was invented by ________ in 1830.
(A) Edwin Lutyens (B) Edwin Budding
(C) J.B. Olcoth (D) William Robinson

846. The book 'Gardens Through the Ages' is authored by
(a) M.S. Randhawa (b) G.S. Randhawa
(c) B.P. Pal (d) T.K. Bose

847. Which of the following trees is effective against CO (carbon monoxide) in the atmosphere?
(A) *Chlorophytum clatum* (B) *Rhapis excelsa*
(C) *Ficus benzamine* (D) *Trachypus fortune*

848. Which of the following is not native to India?
(A) *Ficus religiosa* (B) *Putranjiva roxuburghii*
(C) *Sterculia diversifolia* (D) *Alstonia scholaris*

849. The perpetual flowering _______ is known as father of modern roses.
(A) Parson's Pink China Rose (B) Hume's Blush
(C) Parks Tea China Rose (D) Slater's Crimson China Rose

850. Snap dragon belongs to which family?
(A) Balsaminaceae (B) Malvaceae
(C) Nyctaginaceae (D) Scrophularaceae

851. The following species of Jasmin produces yellow flowers except
(A) *J. primulinum* (B) *J. rex*
(C) *J. humile* (D) *J. Pubigerum*

852. Bird of Paradise, *Sterilitzia reginae* is native to
(A) Mexico (B) South Africa
(C) Sri Lanka (D) Thailand

853. Statice (Limonium) belongs to
(A) Nyctaginaceae (B) Apocynaceae
(C) Lathyraceae (D) Plumbaginaceae

854. Which of the following is scented and early blooming annual?
(A) Petunia (B) Sweet Alyssum
(C) Antirrhinum (D) Coreopsis

855. The climber, *Campsis radicans* produces flowers in ______ colour.
(a) White (b) Orange
(c) Yellow (d) Red

856. Cytoplasmic male sterility is found in
(a) Antirrhinum (b) Salvia
(c) Petunia (d) Marigold

857. The gene which is control the fragrance?
(a) Oligo (b) Aditive
(c) Mono (d) Poly

858. Which of the following herbaceous perennial is good for cut foliage
(a) Adinatum (b) Jacaranda
(c) Cycas (d) Petunia

859. In which of the following groups of ornamental plants, ethylene is involved in induction of flowering?

(a) Ferns (b) Cacti and succulents
(c) Bromeliads (d) Palms and cycads

860. Which fern is most popular cut green?

(a) Pteris (b) Rumhora
(c) Polypodium (d) Adiantum

861. Which of the following trees produces silvery leaves that resemble fern leaves?

(a) *Grevillea robusta* (b) *Melia dubia*
(c) *Jacaranda acutifolia* (d) *Lagerstroemia speciosa*

862. Arrangement of fruits and flowers is known as ....

(a) Ikenobo (b) Ikebana
(c) Morimono (d) Moribana

863. Mutations that occur due to base pair additions and deletions are collectively referred as

(a) Transitions (b) Transversions
(c) Point mutations (d) Frame-shift mutations

864. Which of the following weedy grass is used as an ornamental potted plant?

(a) *Cyperus alternifolium* (b) *Panicum maximum*
(c) *Echinocola colona* (d) *Pennisetum pedicellatum*

865. Wreaths are laid down to pay respect to.........

(a) Marriage function (b) Departed soul
(c) Birthday (d) Welcoming a guest

866. Which of the following greenhouse styles do you recommend for hilly areas?

(a) Lean-to (b) Even-span
(c) Uneven-span (d) Ridge-and-furrow

867. The ideal pH range for soil based and soilless substrates for greenhouse cultivation of a majority of the flower crops is

(a) 4.5 to 5.8 and 4.5 to 5.5, respectively
(b) 6.2 to 6.8 and 5.4 to 6.0, respectively
(c) 7.0 to 7.5 and 7.0 to 7.1, respectively
(d) 7.5 to 8.5 and 5.0 to 5.4, respectively

868. Hydro seedling is one of the latest methods for establishing

(a) Plug plants in portrays in huge greenhouses
(b) Lawn in vast areas
(c) Avenue trees on highways
(d) Hedges over long stretch

869. Which is climbing and evergreen genus of bougainvillea?
(a) *B. spectabilis* (b) *B.* glabra
(c) *B. buttiana* (d) *B. spectoglabra*

870. A popular lawn grass species *Zoysia japonica* is commonly known as ....
(a) Blue-green grass (b) Doob grass
(c) Japanese grass (d) Nut grass

871. Which of the following succulents looks like pebbles around them?
(a) *Pachypodium lamerei* (b) *Adenium obessum*
(c) *Lithops hookeri* (d) *Beaucarnia recuruata*

872. Which of the following cacti is almost spineless?
(a) Zygocactus (b) Opuntia
(c) Haworthia (d) Echinopsis

873. Zen gardens are also known as
(a) Mughal gardens (b) English gardens
(c) Japanese gardens (d) Italian gardens

874. Which of the following tall shrubs is widely propagated by seeds?
(a) *Acalypha hispida* (b) *Allamanda cathartica*
(c) *Putranjiva roxturghii* (d) *Mussaenda erythrophylla*

875. Which of the following indoor plants is poisonous?
(a) Cordyline (b) Chlorophytum
(c) Dieffenbachia (d) Draceana

876. In which condition flowers does not open
(a) Cleistogamy (b) Chasmogamy
(c) Dichogamy (d) Herkogamy

878. Which of the following breeding method is used for transfer of disease resistance?
(a) Pedigree method (b) Bulk method
(c) Mutation breeding (d) Backcross method

879. Best time for Bonsai planting is ....
(A) Nov-Dec (B) July-Aug
(C) Feb-March (D) Both B & C

880. Best repotting time for evergreen Bonsai is ....
(A) Dec-Jan (B) July-Aug
(C) March-April (D) Any of above

881. Head quarter of International cut flower grower association is located at ....
(A) Brazil (B) USA
(C) Italy (D) Canada

882. Head quarter of International bulb research centre is located at ....

(A) Canada (B) Netherland
(C) Poland (D) Kenya

883. What are vulnerable objectives of breeding ornamentals in 21st century?

(A) Host plant resistance to insect-pests and diseases.
(B) Efficient utilization of water and nutrients.
(C) Improved post-harvest life.
(D) All of above

884. What are most exciting breeding tools to develop cultivars in the 21st century?

(A) Somaclonal hybridization
(B) Somaclonal variations
(C) Transformation
(D) All of above

885. Bougainvillea cultivar Chitra is developed through cross between ....

(A) Dr BP Pal x Princess Margret Rose
(B) Tetra Mrs. Mcclean x Dr BP Pal
(C) Dr BP Pal x Mrs. Chico
(D) Dr BP Pal x Jennifer Fernie

886. Which hybrid developed through cross between Dr BP Pal x Mrs. Chico?

(A) Mary Palmer Special (B) Begum Sikander
(C) Wazid Ali Shah (D) None of these

887. Archana variety of Bougainvillea is developed through ....

(A) Back cross (B) Mutation
(C) Hybridization (D) Pure line

888. What are important spontaneous mutated cultivars of Bougainvillea?

(A) Archana (B) Shweta
(C) Surekha (D) All of these

889. What are important double bracts bougainvillea cultivars?

(A) Cherry Blossom (B) Mahara
(C) Rosevilles Delight (D) All of these

890. Who is not associated with Chrysanthemum?

(A) SK Datta (B) TN Khoshoo
(C) MN Gupta (D) None of these

891. Who was associated with Jasmine?

(A) Bhupal Rao (B) MN Gupta
(C) SN Zadoo (D) All of these

892. R.N. Bhatt was associated with ....
(A) Bougainvillea (B) Hibiscus
(C) Tuberose (D) Gladiolus

893. Which are leading institutions in breeding hibiscus?
(A) IIHR (B) TNAU
(C) Lalbagh garden (D) All of above

894. SPS Raghava was associated with ....
(A) Bougainvillea (B) Hibiscus
(C) Tuberose (D) Gladiolus

895. Swami Vinayananda was associated with ....
(B) Dahlia (B) Hibiscus
(C) Marigold (D) Gerbera

896. Which is the leading institute for Dahlia breeding?
(A) IIHR (B) IARI
(C) BSI Kolkata (D) FRI

897. Which of the following is not an African Marigold variety?
(A) Pusa Arpita (B) Cracker Jack
(C) Pusa Narangi Gainda (D) Crown of Gold

898. Which of the following is a French Marigold variety?
(A) Spun Gold (B) Climax
(C) Star of India (D) Pusa Basanti Gainda

899. Colour magic is an cultivar of ....
(A) African marigold (B) French marigold
(C) Dahlia (D) Tuberose

900. A lawn can be described as ....
(A) A land covered with lawn grass
(B) Green carpet
(C) A piece of land in a garden
(D) All of these

901. Which of the following is known as miniature garden?
(A) Mughal (B) Japanese
(C) Italian (D) Persian

902. Floribunda rose is a cross between ....
(A) Hybrid Teas × Perpetual Polianthas
(B) Tea Rose × Perpetual Polianthas
(C) Hybrid Perpetual × Tea Rose
(D) Perpetual Polianthas × Hybrid Perpetual

903. Which of the following is not used for flower preservation?
(A) Sugar (B) Polyols
(C) KCL (D) H2SO4

904. Lawn area in home garden should not be more than ....
(A) 30% (B) 20%
(C) 50% (D) 40 %

905. Spores are used to propagate ....
(A) Mosses (B) Asparagus
(C) Ferns (D) Bromeliads

906. Which of the following is a mutant variety of Jasmine?
(A) Pitchi (B) Bio- 13
(C) MTU- 8 (D) Co- 9

907. Shape of Hogarth course line is ....
(A) Strait lines (B) Serpentine curved lines
(C) Zigzag lines (D) curved lines

908. To improving vase life of cut flowers it should kept in ....
(A) Red or blue, 2000 lux or more
(B) Red or blue, 500 lux or more
(C) Red or violet 1500 lux or more
(D) Red or violet 1000 lux or more

909. Silver thiosulphate is not used for ....
(A) Rose (B) Carnation
(C) Tulip (D) Lily

910. Which of the following is day neutral plant?
(A) Amaranthus (B) Gompherena
(C) Aloe (D) Gulmohar

911. Swarn Rekha and Vaibhav are important cultivar of which flower?
(A) Lilly (B) Tuberose
(C) Gladiolus (D) Dahlia

912. Dry flower arrangement is known as ....
(A) Ikebana (B) Pot – Porris
(C) Flower bouquet (D) Both a and b

913. Which of the following is used for drying of Ornamentals?
(A) Corn granules (B) Silica gel
(C) Both a and b (D) None of above

914. Botanical of Bird net's fern is ....

(A) *Asplenium nidus* (B) *Matteuccia struthiopteris*

(C) *Pteridium aquilinum* (D) *Adiantum capillus-veneris*

915. Herbaceous term was first used by ....

(A) G. Jekyell (B) William Robinson

(C) H. Repton (D) W. Kent

916. Coco peat mostly used for ....

(A) Seed germination tray (B) Preparation of nursery bed

(C) It is applied in barren land (D) It is used in kitchen garden

917. Black spot disease in rose is caused by ....

(A) Diplocarpon (B) Alternaria

(C) Xanthomonas (D) Diploma

918. In water garden kachha pools are ....

(A) Informal (B) Formal

(C) Modern style (D) Free style

919. Style of pakka pools in water garden is ....

(A) Informal (B) Formal

(C) Modern style (D) Free style

920. Marigold should not be planted near ....

(A) Oilseed crops (B) Legume crops

(C) Cereal crops (D) Fiber crops

921. Which of the following is not a conifer?

(A) Cypresses (B) Casuarina

(C) Abies (D) Thuja

922. Which of the following wire is best for bonsai making?

(A) Copper (B) Iron

(C) Aluminium (D) Fiber

923. Quilling florets disorder found in ....

(A) Lilly (B) Chrysanthemum

(C) Gladiolus (D) Dahlia

924. Propagation material for gladiolus is ....

(A) Cormels (B) Corm

(C) Bulb (D) Both a and b

925. What is the botanical name of Indian Juniper?

(A) *Juniperus macropoda* (B) *Juniperus indica*

(C) *Juniperus recurba* (D) *Juniperus communis*

926. Curry leaf originated from ....
(A) Indonesia (B) India
(C) USA (D) China

927. Which of the following is a pest of dried flowers?
(A) Termites (B) Silver moth
(C) Silver fish (D) Book caterpillar

928. Bull head in rose caused by ....
(A) Hoppers (B) Thrips
(C) Aphid (D) White fly

929. Rambagh was developed by ....
(A) Babur (B) Akbar
(C) Humayun (D) Shah Jahan

930. NHB is located at ....
(A) Ludhiana (B) New Delhi
(C) Gurgaon (D) Pusa

931. Which of the does not appear flowers ....
(A) Dieffenbachia (B) Archid
(C) Hyacinth (D) Peony

932. Winter queen is cultivar of ....
(A) Lilly (B) Gerbera
(C) Gladiolu (D) Dahlia

933. Flowering plants comes under group ....
(A) Gymnospermae (B) Magnoliophyta
(C) Pteridophyta (D) Ginkhophyta

934. Which of the following is known as "Bean of India"?
(A) Daffodils (B) Lotus
(C) Rose (D) Gladiolus

935. Nagerie is a type of ....
(A) Garden (B) Ikebana
(C) Bonsai art (D) Trophy

936. Which of the following is a characteristic of dry flower?
(A) Longevity (B) Nobility
(C) Aesthetics (D) All of the above

937. MDU-1 is a variety of which flower?
(A) Daffodils (B) Chrysanthemum
(C) Rose (D) Gladiolus

938. Petunia is an ………… Plant

(A) Cross pollinated (B) Often cross pollinated

(C) Self-pollinated (D) None of the above

939. Botanical name of goose grass is ……….

(A) *Stenotaphrum secundatum* (B) *Cynodon dactylon*

(C) *Eleusine indica* (D) *Festuca fratensis*

940. Which of the following has important palace in home garden?

(A) Christmas tree (B) Lawn

(C) Bonsai (D) Rose

941. Which of the following is ideal for lawn?

(A) Ups and downs (B) Plain

(C) Sloppy (D) All of the above

942. To improve vase life of cut flowers ………… Is used

(A) 8- HQC (B) Cycocel

(C) Gibberellin (D) Glucose

943. Botanical name of lurk spur is …………

(A) *Althea rosea* *(B) Delphinium ajacis*

(B) Iberis amara *(D) Lupinus tricolor*

944. Which of the following is not a winter season annual?

(A) Sweet Sultan (B) Varbina

(C) Corn flower (D) Sweet alyssum

945. Jasmine is the national flower of …………

(A) Indonesia (B) Philippines

(C) Pakistan (D) All of these

946. Dwarf protective hedge have ………. meter height

(A) 2 (B) 2-3

(C) 1 (D) 3-4

947. Which of the following is a succulent?

(A) Sea onion (B) Yucca

(C) Opuntia (D) All of these

948. Xanthophyll is extracted from which part of marigold?

(A) Leaves of marigold (B) White petals of marigold

(C) Yellow perals of marigold (D) Seeds of marigold

949. Shalimar garden is located at …………

(A) Srinagar (B) Lahore

(C) Islamabad (D) Jammu

950. Lotus is a national flower of which country?
(A) India and Sri Lanka (B) India and Vietnam
(C) India and Bangladesh (D) India and Burma

951. Crossandra belong to family ....
(A) Apocyanaceae (B) Apiaceae
(C) Acanthaceae (D) Iridaceae

952. Mandor garden located at ....
(A) Jaipur (B) Jodhpur
(C) Gurgaon (D) Haryana

953. National botanical garden is situated at ....
(A) Varanasi (B) Lucknow
(C) Kolkata (D) Ooty

954. Heart of garden is known as ....
(A) Rockery (B) Lawn
(C) Trophy (D) Topiary

956. Tushar is a .... Colour variety of rose
(A) Yellow (B) White
(C) Pink (D) Red

957. Crimson is a .... Colour variety of rose
(A) Yellow (B) White
(C) Red (D) Pink

958. Friendship is a variety of ....
(A) Daffodils (B) Gladiolus
(C) Rose (D) Chrysanthemum

959. Plant form of kochia is ....
(A) Columnar (B) Informal
(C) Conical (D) Globular

960. Which chemical is used for defoliation of rose?
(A) Urea (B) Copper sulphate
(C) Auxin (D) GA

961. Chromosome number of African marigold is 2n = ...
(A) 24 (B) 48
(C) 12 (D) 36

962. Which of the following climber is suitable for screening purpose?
(A) Passiflora (B) Thunbergia
(C) *Antigonon leptopus* (D) All of these

963. Which of the following is a foliage climber?
(A) Aristolochia (B) *Quisqualis indica*
(C) Monstera (D) Passiflora

964. Antigonon leptopus is belong to family ....
(A) Balsaminaceae (B) Polygonaceae
(C) Nyctaginaceae (D) Scrophularaceae

965. *Adenocalymma allicea* is also known as ....
(A) Garlic vine (B) Coral vine
(C) Onion vine (D) Grape vine

966. Cat's claw belong to family ....
(A) Balsaminaceae (B) Bignoniaceae
(C) Nyctaginaceae (D) Scrophularaceae

967. Botanical name of Glory lily is ....
(A) *Clitoria ternate* (B) *Gloriosa superba*
(C) *Combretum comosum* (D) None of above

968. Cup and saucer plant belong to which family?
(A) Balsaminaceae (B) Convonvoluceae
(C) Varbanaceae (D) None of above

969. Floss flower belong to which family?
(A) Compositae (B) Malvaceae
(C) Amaranthaceae (D) Croyophyllaceae

970. Cock's comb propagated by ....
(A) Seeds (B) Cuttings
(C) Tissue culture (D) Layering

971. Gaillardia is also known as ....
(A) Floss flower (B) Blanket flower
(C) Pot marigold (D) Duck flower

972. Pot marigold is an another name of..........
(A) Calendula (B) Marigold
(C) Annual chrysanthemum (D) Orchid

973. Phlox is propagated by ....
(A) Division (B) Suckers
(C) Seeds (D) Slips

974. Salvia splendens belong to family ....
(A) Liliaceae (B) Labiateae
(C) Leguminuceae (D) Violaceae

975. Periwinkle belong to which family ....

(A) Acantheceae (B) Apocynaceae

(C) Malveceae (D) None of above

976. Which of the following is suitable for rockery purpose?

(A) Nasturtium (B) Verbena

(C) phlox (D) All of these

977. Which of the following is a cycad?

(A) Adiantum (B) Zamia

(C) Pritchardia (D) *Caryota urens*

978. Golden shower belong to family ....

(A) Malveceae (B) Leguminosae

(C) Guttiferae (D) Boraginaceae

979. Croton comes under which family?

(A) Euphorbiaceae (B) Tiliaceae

(C) Guttiferae (D) Malveceae

980. Alternate anthers are commonly ....

(A) Fertile (B) Infertile

(C) Both a & b (D) None of above

981. *Acalypha hispida* is commonly grown for ....

(A) Flowers (B) Spike

(C) Leaves (D) All of above

982. *Michilia champaka* have .... Colour flowers

(A) Light yellow (B) Dark yellow

(C) White (D) Red

983. Which of the following is the symbol of youth and life?

(A) White kachnar (B) Water

(C) Marigold (D) River

984. Maulsari belong to family ....

(A) Sapoteceae (B) Tiliaceae

(C) Bignoniaceae (D) Termindeceae

985. *Casia nodosa* appear .... colour flowers

(A) Red (B) Pink

(C) Yellow (D) Orange

986. Which of the following garden known as genesis of gardening?

(A) Eden (B) Osaka

(C) Hampshire (D) Brindavan

987. Osaka garden located at ....

(A) Japan (B) Australia

(C) USA (D) Italy

988. Which of the following garden was not described by Vatsyana?

(A) Parmadodyan (B) Rock garden

(C) Nandavana (D) Both b & c

989. By Mughals which plant grown as a symbol of death and eternity?

(A) Kachnar (B) Pomegranate

990. Roshanara park New Delhi was developed by ....

(A) Sir Edwin Lutyens (B) Prof. K. Mori

(C) Prof. H. Mori (D) Nara

991. Roshanara park is an .... style garden

(A) Mughal (B) Japanese

(C) English (D) Persian

992. In Japanese language sand garden known as ....

(A) Roji-Niwa (B) Royanji

(C) Hira-Niwa (D) Uchi-roji

993. Sim's park is situated at ....

(A) Coonoor (B) Kodaikanal

(C) Ooty (D) Madurai

994. Bryant park is located at ....

(A) Coonoor (B) Kodaikanal

(C) Ooty (D) Madurai

995. Indian botanic garden is situated opposite to which river?

(A) Ganga (B) Hoogly

(C) Yamuna (D) Brahmaputra

996. Which of the following garden now known as NBRI?

(A) Roshanara park (B) Sikander bagh

(C) Wajid Ali Bagh (D) None of above

997. Stone lanterns are important feature of which style garden?

(A) Mughal (B) Japanese

(C) English (D) Persian

998. Which of the following is not comes under formal garden?

(A) Mughal (B) French

(C) English (D) Italian

999. Sayaji park is located at ....

(A) Pune (B) Baroda

(C) Mumbai (D) Goa

1000. Leading Jasmine perfume producing country?

(A) France (B) Egypt

(C) Bulgaria (D) The Netherland

1001. Which is a blue colour variety of Rose?

(A) Abhisharka (B) Priya

(C) Samba (D) Pusa Christina

1002. First pinching in Carnation is done at the stage of?

(A) 4TH pair leaf (B) 8TH pair leaf

(C) 6TH pair leaf (D) All of above

1003. Petal bum in Chrysanthemum is due to deficiency of?

(A) Boron (B) Calcium

(C) Cobalt (D) Nitrogen

1004. Nugget is a triploid variety of which flower?

(A) Rose (B) Carnation

(C) Marigold (D) ladiolus

1005. Exalbuminous seeds are found in which flower?

(A) China aster (B) Orchid

(C) Marigold (D) Tuberose

1006. Leading flowers seed producing state in India?

(A) Haryana (B) Tamil Nadu

(C) West Bengal (D) Punjab

1007. Pseudo bulbs are found in which flower crop?

(A) Tuberose (B) Canna

(C) Orchid (D) All of these

1008. Gladiolus is propagated by?

(A) Bulb (B) Tuber

(C) Corm (D) Rhizome

1009. Gladiolus cultivar Shobha is developed through which method?

(A) Hybridization (B) Selection

(C) Mutation (D) None of these

1010. Dahlia is propagated through?

(A) Tuber (B) Rhizome

(C) Cuttings (D) Tuberous root

1011. Swami vinyanand was related to which flower crop?
(A) Dahlia (B) Gladiolus
(C) Chrysanthemum (D) Hibiscus

1012. Scoring is important practice of which flower crop?
(A) Tuberose (B) Gladiolus
(C) Dahlia (D) All of these

1013. Planting time for spring flowering crop?
(A) January (B) December
(C) November (D) August

1014. Mayur variety of Gladiolus is released from?
(A) IIHR (B) NBRI
(C) IARI (D) AU Punjab

1015. How many chromosome numbers are there in Gladiolus?
(A) 30 (B) 40
(C) 60 (D) 16

1016. Negative geotropism in Gladiolus is occur during?
(A) Vegetative period (B) Storage
(C) Transporting (D) Flowering

1017 Breaking virus in tulip is transferred through?
(A) White fly (B) Thrips
(C) Aphids (D) Honey bee

1018. International registration authority of Rose is situated at?
(A) USA (B) China
(C) New Delhi (D) Bulgaria

1019. Which state produce highest cut flower in India?
(A) Karnataka (B) Tamil Nadu
(C) Rajasthan (D) West Bengal

1020. Which of the following flower is known as barbeton daisy?
(A) Gladiolus (B) Gerbera
(C) Tulip (D) Pansy

1021. Fluoride injury mostly found in which flower crop?
(A) Tuberose (B) Gladiolus
(C) Rose (D) All of these

1022. Tinting is an important practice of which flower crop?
(A) Rose (B) Carnation
(C) Hibiscus (D) Jasmine

1023. Which of the following bulbous crop cover highest area in India?
(A) Tuberose (B) Dahlia
(C) Gladiolus (D) Canna

1024. Which of the following is a thorn less variety of Rose?
(A) Suchitra (B) Grand Gala
(C) Pusa Mohit (D) All of these

1025. African Marigold is a symbol of?
(A) Vulgar mind (B) Jealousy and Sorrow
(C) Self-Steam (D) Purity

1026. Optimum spacing for hybrid tea Rose?
(A) 60 cm (B) 90 cm
(C) 75 cm (D) 65 cm

1027. Which of the following is a thorn less species of Rose?
(A) *R. pendulina* (B) *R. bourboriana*
(C) *R. canina* (D) Both A & B

1028. Which of the following is a multi-coloured mutant variety of Rose?
(A) Abhisharka (B) Madhos
(C) Priya (D) All of these

1029. Haldi Ghati is a variety of which flower?
(A) Rose (B) Dahlia
(C) Chrysanthemum (D) Carnation

1030. Which of the following flower is known as Rose of Indies?
(A) Hibiscus (B) China aster
(C) Marigold (D) Dahlia

1031. China aster is propagated through?
(A) Cuttings (B) Micro propagation
(C) Seeds (D) Rhizome

1032. Which of the following orchid is monopodial?
(A) Dendrobium (B) Vanda
(C) Cattleya (D) Bulbophyllum

1033. Which type of fruits are present in Rose?
(A) Capsule (B) Berry
(C) Nut (D) None of these

1034. Maximum oil recovery is done by which species?
(A) J. sambac (B) J. Grandiflorum
(C) J. auriculatum (D) J. multiflorum

1035. Mickey mouse is a variety of which flower?
(A) Carnation (B) Orchid
(C) Anthurium (D) Portulca

1036. Chrysanthemum cultivar Money Maker is developed through?
(A) Biotechnology (B) Selection
(C) Hybridization (D) Mutation

1037. Which of the following first variety of Floribunda Rose?
(A) La France (B) Rodhatte
(C) Buccaner (D) Grand gala

1038. How much seeds of Marigold are required for one hectare?
(A) 1-1.5 Kg (B) 4-5 Kg
(C) 500 gm (D) 300 gm

1039. Which of the following largest genera of Orchid?
(A) Bulbophyllum (B) Dendrobium
(C) Vanda (D) Oncidium

1040. Which of the following storage disorder of Rose?
(A) Bull head (B) Bent neck
(C) Sleepiness (D) None of these

1041. Which of the following Rose variety have great demand in cut flower market?
(A) Raktagandha (B) Sindhoor
(C) Arjun (D) All of these

1042. Which of the following is an off season variety of Tuberose?
(A) Rajat Rekha (B) Shringar
(C) Suvashini (D) Pearl Double

1043. Nugget variety of Marigold was developed from?
(A) Egypt (B) USA
(C) India (D) Netherland

1044. Rose variety Pusa Mohini in India introduce from?
(A) USA (B) China
(C) Germany (D) Italy

1045. Inflorescence type of Chrysanthemum?
(A) Spike (B) Spadix
(C) Capitulum (D) None of these

1046. What is mean of pot mums?
(A) 1 Plant in 1 Pot (B) 1 Species in 1 Pot
(C) 1 Species in >1 Pot (D) 1 Plant in >1 Pot

1047. Who was introduce Damask Rose in India?
(A) Baber (B) British
(C) Araya (D) B.P. Pal

1048. Scoring is an important operation of which flower
(A) Tulip (B) Gladiolus
(C) Hibiscus (D) Rose

1049. Foliage Nematode is a serious problem of which flower?
(A) Tuberose (B) Gladiolus
(C) Orchid (D) All of these

1050. Which of the following type pollination found in Orchid?
(A) Animophilous (B) Hydrophilous
(C) Entomophilous (D) Ornithophilous

1051. Economic part of Anthurium?
(A) Leaves (B) Spathe
(C) Whole plant (D) Spadix

1052. Which of the following flower is an indicator of fluoride?
(A) Chandni (B) Gladiolus
(C) Tuberose (D) Kamini

1053. Disbudding in Dahlia is done at which stage?
(A) Bud appearance stage (B) Pea stage
(C) Bud opening stage (D) Flower opening stage

1054. Stem break is a post-harvest disorder of which flower?
(A) Dahlia (B) Gerbera
(C) Tuberose (D) Gladiolus

1055. Lotus is propagated through which method?
(A) Suckers (B) Seeds
(C) Cuttings (D) All of these

1056. Dr. B.P. Pal is a variety of which flower?
(A) Bougainvillea (B) Rose
(C) Chrysanthemum (D) Gerbera

1057. Which of the following Rose is mainly used for gulkhand?
(A) Damask Rose (B) Chinese Rose
(C) Edward Rose (D) Dog Rose

1058. Colour break in Anthurium spathe is due to deficiency of?
(A) Calcium (B) Boron
(C) Nitrogen (D) Phosphorus

1059. Which of the following pollination agent in Petunia?
(A) Honey bees (B) Air
(C) Water (D) Birds

1060. Yellow colour is extracted from which Rose species?
(A) *R. foetida* (B) *R. canina*
(C) *R. galica* (D) All of these

1061. Shri Bajrang Bahadur Singh is related to which flower?
(A) Gladiolus (B) Chrysanthemum
(C) Jasmine (D) Gerbera

1062. Wilt of Rose is caused by?
(A) Virus (B) Viroid
(C) Fungus (D) Bacteria

1063. Apply artificial colour in Carnation is known as?
(A) Pulsing (B) Tinting
(C) Precooling (D) Hilling

1064. Madhosh variety of Rose is developed through?
(A) Introduction (B) Mutation
(C) Selection (D) Demonstration

1065. Arka Archana is a variety of which flower?
(A) Zinnia (B) Dahlia
(C) China aster (D) Calendula

1066. Propagation in Bougainvillea done through?
(A) Semi hard wood cutting (B) Hard wood cutting
(C) Layering (D) Soft wood cutting

1067. Bronze colour foliage in Chrysanthemum is due to deficiency of?
(A) Boron (B) Led
(C) Calcium (D) Potash

1068. Topple & bud rot in Lily is due to deficiency of?
(A) Calcium (B) Aluminium
(C) Iron (D) Cobalt

1069. India's share in the global flower market?
(A) 1 % (B) 0.6 %
(C) 2 % (D) 0.8 %

1070. Chrysanthemum in Hindi is known as?
(A) Guldaudi (B) Genda
(C) Ghal (D) Gulab

1071. Bharat Sundari is an important cultivar of which flower?
(A) Hibiscus (B) Marigold
(C) Tulip (D) Dahlia

1072. Banjaran cultivar of Rose is related to which type of Rose?
(A) Hybrid tea (B) Tea Rose
(C) Floribundas (D) None of these

1073. Queen Elizabeth is a cultivar of which flower?
(A) Chrysanthemum (B) Rose
(C) Carnation (D) Lily

1074. Birbal Sahnii is a variety of?
(A) Chrysanthemum (B) Dahlia
(C) Marigold (D) Tulip

1075. African Marigold is an?
(A) Triploid (B) Diploid
(C) Aneuploidy (D) Tetraploid

1076. Mostly the cultivars of Chrysanthemum are?
(A) Short day (B) Long day
(C) Day neutral (D) Semi short day

1077. Which of the following practice is followed for large sized flowers?
(A) Pinching (B) Dishooting
(C) Disbudding (D) All of these

1078. Which of the following substance is used for breaking Gladiolus corm dormancy?
(A) ABA (B) GA3
(C) NAA (D) SADH

1079. Shoot tip culture is commonly followed in which flowers?
(A) Carnation (B) Chrysanthemum
(C) Orchid (D) All of these

1080. Thimma is an important cultivar of?
(A) Jasmine (B) Marigold
(C) Bougainvillea (D) Chrysanthemum

1081. Spanish Jasmine is known as ....
(A) *J. grandiflorum* (B) *J. sambac*
(C) *J. auriculatium* (D) *J. multiflorum*

1082. Damping off is caused by?
(A) *Pythium* (B) *Rhizoctonia*
(C) *Phytophthora* (D) All of these

1083. Mohini is a which colour cultivar of Rose?
(A) Blue (B) Chocolate
(C) Red (D) Yellow

1084. Most beautiful flower on this globe?
(A) Rose (B) Orchid
(C) Lotus (D) Gladiolus

1085. Jewel Orchids are valued for their beautiful?
(A) Leaves (B) Flower
(C) Roots (D) Stigma

1086. *Eria pusilla* Orchid has how much height?
(A) 1 cm (B) 30 cm
(C) 5 cm (D) 10 cm

1087. Fruit of Rose is known as ....
(A) Jip (B) Hip
(C) Column (D) Bud spot

1088. Best time for budding in Rose in North Indian plains?
(A) October (B) November
(C) January (D) August

1089. Hilly areas are most suitable for growing quality ....flowers
(A) Annual (B) Bulbous
(C) Biennial (D) Perennial

1090. Which is an inter specific hybrid of Bougainvillea?
(A) Begum Sikander (B) Wazi Ali Shah
(C) Chitra (D) All of these

1091. Gladiolus is originated from......?
(A) America (B) South Africa

1092. Mohini is a .... cultivar of Rose
(A) Diploid (B) Triploid
(C) Aneuploid (D) Haploid

1093. Optimum storage temperature for gladiolus corm is .... ºC
(A) 0 – 1 ºC (B) 2 – 5 ºC
(C) 4 – 7 º (D) 10 -12 ºC

1094. Which of the following is not a fragrant variety of Gladiolus?
(A) Lucky star (B) Jimmy boy
(C) Sager (D) All of these

1095. Punjab gold is a variety of ....
(A) Marigold (B) Chrysanthemum
(C) Jasmine (D) Gladiolus

1096. Arthur Sim Carnation is highly resistance to ....
(A) Wilt (B) Rust
(C) Stem rot (D) All of these

1097. Who was started tissue culture in Orchid?
(A) Lysenko (B) Shull
(C) Morel (D) S. Srivastava

1098. Calyx splitting in Carnation is due to ....
(A) Genetic factor (B) Environmental factor
(C) Nutritional factor (D) All of these

1099. Most serious pest of Carnation........
(A) Aphids (B) Thrips
(C) Red spider mite (D) White fly

1100. Gulkhand is prepared by using petals and sugar in ratio of ......
(A) 1:1 (B) 1:2
(C) 2:1 (D) 1:3

1101. Chemical defoliation in Rose is done by ....
(A) Urea (B) $CuSo_4$
(C) GA3 (D) ABA

1102. Which colour Roses are most popular?
(A) Red (B) Pink
(C) Yellow (D) Blue

1103. In cut Roses minimum vase life should be .... days
(A) 10 (B) 12
(C) 8 (D) 14

1104. Which of the following is not a sympodial orchid?
(A) Cattleya (B) Vanda
(C) Epidendrum (D) All of above

1105. Orchid plant looking healthy but not produce flowers it is due to improper ......
(A) Temperature (B) Light
(C) Nutritional (D) All of above

1106. Generally orchids require .... foot candles light intensity
(A) 1000-2000 (B) 2000-2500
(C) 3000-6000 (D) 4000-5000

1107. According to photoperiod orchid is a .... plant

(A) Long day (B) Day neutral

(C) Short day (D) CAM plant

1108. Orchids require .... % humidity for well grown

(A) 50-60 (B) 40-60

(C) 60-80 (D) 70-80

1109. Orchid flower should be stored at .... °C temperature

(A) 0-2 (B) 4-7

(C) 10 (D) 10-12

1110. Anthurium is a .... word

(A) French (B) Greek

(C) Arabic (D) None of above

1111. Largest producer of anthurium is .... in the world

(A) Germany (B) USA

(C) Netherland (D) France

1112. Anthurium is a .... plant

(A) Protoandry (B) Protogynous

(C) Both a & b (D) Nether a nor b

1113. Fruit type of anthurium is ....

(A) Nut (B) Berry

(C) Stone (D) Pom

1114. Anthurium belong to family ....

(A) Acantheseae (B) Araceae

(C) Rosaceae (D) Amarantheceae

1115. Around .... cultivated species of anthurium found in the world

(A) 15 (B) 50

(C) 100-110 (D) 4-5

1116. Which of following is not a flowering species of anthurium?

(A) *Ànthurium andreanum* (B) *A. scherzerianum*

(C) *Anthurium leuconerum* (D) None of above

1117. Which of following is not a foliage species of anthurium?

(A) *A. corrugatum* (B) *A. crystallinum*

(C) *A. robustum* (D) *A. clarinervium*

1118. Which of the following species is known as oil cloth flower?

(A) *Anthurium andreanum* (B) *A. scherzerianum*

(C) *Anthurium leuconerum* (D) None of above

1119. Which of the following is known as flame plant?
(A) *Anthurium andreanum* (B) *A. scherzerianum*
(C) *Anthurium leuconerum* (D) *A. corrugatum*

1120. How many chromosomes are present in anthurium?
(A) 26 (B) 30
(C) 60 (D) 40

1121. Up right cupped spathe present in which type anthurium?
(A) Obake (B) Standard
(C) Tulip (D) All of above

1122. Which of the following is not a standard type variety of anthurium?
(A) Ozaki (B) Kalapana
(C) Nitta (D) Midori

1123. Which of the following is not a obake type variety?
(A) Micky mouse (B) Rainbow
(C) Kozohara (D) Anuenue

1124. Required light intensity for anthurium cultivation is .... lux
(A) 1000-2000 (B) 20000-25000
(C) 5000-6000 (D) 8000-10000

1125. Required humidity level for anthurium cultivation is .... %
(A) 50-85 (B) 50-60
(C) 40-60 (D) 75-80

1126. Growing media for anthurium should have .... P[H]
(A) 4.5-5.5 (B) 5
(C) 5-6.5 (D) 6-7

1127. Anthurium spathe is the modification of ....
(A) Leaf (B) Stalk
(C) Root (D) Candle

1128. Which of the following spacing adopted for commercial cultivation of anthurium?
(A) 45X45 (B) 30X45
(C) 25X30 (D) 20X30

1129. After .... months of anthurium planting we can harvest flowers
(A) 6-8 (B) 8-9
(C) 10-12 (D) 4-5

1130. Anthurium can be easily stored at .... °C temperature
(A) 4-5 (B) 13
(C) 5-7 (D) 2

1131. *Phytophthora parasitica* cause which disease in anthurium?
(A) Soft rot (B) Root rot
(C) Flower blight (D) Anthracnose

1132. Which of the following is known as Transvaal daisy?
(A) Gladiolus (B) Gerbera
(C) Dahlia (D) Lotus

1133. Gerbera belong to which family?
(A) Araceae (B) Asteraceae
(C) Acantheceae (D) Amarantheceae

1134. Which of the following is only a cultivated species of gerbera?
(A) *Gerbera jamesonii* (B) *Gerbera asplenifolia*
(C) *Gerbera aurantiaca* (D) *Gerbera viridifolia*

1135. Optimum required temperature is .... °C for gerbera seed germination
(A) 10-12 (B) 15-20
(C) 12-14 (D) 20-22

1136. Optimum required humidity for gerbera cultivation is .... %
(A) 60-70 (B) 70-75
(C) 75-80 (D) 80-85

1137. Optimum required temperature is .... C for flowering in gerbera
(A) 20 (B) 23
(C) 25 (D) 15

1138. Double faced flower in gerbera is caused by ....
(A) Genetic factor (B) Nutrient imbalance
(C) Environmental factor (D) All of these

1139. First flower produce in gerbera .... weeks after planting
(A) 6-7 (B) 7-8
(C) 10-12 (D) 5-6

1140. Carnation belong to which family?
(A) Caryophyllaceae (B) Araceae
(C) Asteraceae (D) Amarantheceae

1141. Carnation flower is also known as ....
(A) Devine flower (B) Clove flower
(C) Gilly flower (D) All of above

1142. Candy is the variety of which flower?
(A) Chrysanthemum (B) Carnation
(C) Dahlia (D) Gerbera

1143. Roma is a red colour variety of ....
(A) Chrysanthemum (B) Carnation
(C) Dahlia (D) Gerbera

1144. For good root development cuttings will take ....
(A) 10 days (B) 21 days
(C) 15 days (D) 30 days

1145. Which of the following concentration of IBA is used for carnation?
(A) 100 ppm (B) 400 ppm
(C) 500 ppm (D) 700 ppm

1146. Which of the following is a standard type variety of carnation?
(A) Excel (B) Candy white
(C) Rony (D) Silvery Pink

1147. Which variety is the cross of Shringar X Maxican single?
(A) Vaibhav (B) Prajwal
(C) Suvasini (D) Rajat Rekha

1148. How much concrete yield obtained from tuberose flowers?
(A) 0.08 to 0.11 % (B) 0.04-0.5%
(C) 0.25-0.5% (D) 0.5-0.75%

1149. Shape of tuberose seeds is ....
(A) Flat (B) Round
(C) Oblong (D) Nonc of above

1150. Crossandra comes under .... Family
(A) Moraceae (B) Malveceae
(C) Acanthaceae (D) Asteraceae

1151. Orange crossandra comes under which group?
(A) Tetraploid (B) Triploid
(C) Diploid (D) None of above

1152. How many chromosome numbers are there in Delhi crossandra?
(A) 16 (B) 30
(C) 40 (D 20

1153. How many months required for flowering after planting in crossandra?
(A) 4-5 (B) 2-3
(C) 3-4 (D) >5

1154. Optimum yield of Delhi crossandra is ....Kg.
(A) 2400 (B) 2000
(C) 2800 (D) 3000

1155. Rose is a .... word

(A) Latin (B) Greek

(C) French (D) Arabic

1156. Which of the following type rose is suitable for hedge purpose?

(A) Floribunda (B) Miniatures

(C) Button roses (D) All of above

1157. Dried petals of rose is known as ....

(A) Hips (B) Bud

(C) Pankhuri (D) All of above

1158. How many weeks will take from planting to flowering in gerbera?

(A) 7-8 (B) 8-10

(C) 10-11 (D) 5-6

1159. Major cut flower growing states in India

(A) West Bengal (B) Karnataka

(C) AP (D) MP

1160. Which is the bougainvillea multi bracted cultivar?

(A) Lady Mary Baring (B) Cherry blossom

(C) Golden glow (D) All of above

1161. Bougainvillea variety Merry Palmer is developed through ....

(A) Mutation (B) Hybridization

(C) Introduction (D) All of these

1162. Propagation of marigold is done by ....

(A) Seeds (B) Cuttings

(C) Layering (D) All of above

1163. Commercial propagation method of rose is ....

(A) T-Budding (B) Cuttings

(C) Seeds (D) Suckers

1164. Propagation method of dahlia is ....

(A) Leaf cutting (B) Stem cuttings

(C) Root cuttings (D) Tuberous cuttings

1165. Carnation interveinal chlorosis is due to deficiency of ....

(A) Boron (B) Mn

(C) Cu (D) Mg

1166. Black spot in rose is caused by ....

(A) Fungi (B) Bacteria

(C) Nutrient deficiency (D) All of above

1167. Grand gala is a variety of ....
(A) Dahlia (B) Rose
(C) Tuberose (D) None of above

1168. Which of the following is highly affected by mite?
(A) Carnation (B) Gerbera
(D) Lilium (D) None of above

1169. Connecticut king cultivar is related to ....
(A) Asiatic lily (B) Rose
(C) Oriental lily (D) Carnation

1170. China aster is an .... Pollinated crop
(A) Cross (B) Self
(C) Often cross (D) None of these

1171. Star gazer is a variety of ....
(A) Asiatic lily (B) Oriental lily
(C) Tiger lily (D) Torch lily

1172. Lilium is propagated through ....
(A) Suckers (B) Slips
(C) Bulbils (D) Tubers

1173. CMS is found in ....
(A) Petunia (B) Marigold
(C) Salvia (D) Anthirinium

1174. In which of the following mechanism flowers does not open?
(A) Dicogamy (B) Harkogamy
(C) Cleistogamy (D) Chasmogamy

1175. Which of the following method is used for disease resistant development?
(A) Back cross (B) Bulk method
(C) Pure line (D) Pedigree method

1176. International flower grower association head quarter located at ....
(A) USA (B) Australia
(C) Switzerland (D) Japan

1177. International bulb research centre is located at ....
(A) Kenya (B) Netherland
(C) USA (D) Canada

1178. Bougainvillea cultivar Chitra is developed through ....
(A) Mutation (B) Introduction
(C) Hybridization (D) None of above

1179. Archana variety of bougainvillea is developed through ....
(A) Mutation (B) Hybridization
(C) Back cross (D) Pure line

1180. Cherry blossom is a cultivar of ....
(A) Bougainvillea (B) Marigold
(C) Aster (D) Rose

1181. S.K. Dutta is associated with ....
(A) Rose (B) Chrysanthemum
(C) Dahlia (D) Tuberose

1182. MA. Kher is related to which flower crop?
(A) Chrysanthemum (B) Hibiscus
(C) Jasmine (D) Dahlia

1183. Bhupal Rao was associated with ....
(A) Jasmine (B) Rose
(C) Gladiolus (D) Tuberose

1184. R.N. Bhatt was associated with ....
(A) Hibiscus (B) Jasmine
(C) Gladiolus (D) None of above

1185. Which of the following is not the line of 'Ikebana'?
(A) Jushi (B) Earth
(C) Man (D) Heaven

1186. Man in Japanese language is known as...........
(A) Shine (B) Soe
(C) Hikae (D) Jushi

1187. Blazing star or Golden star belong to which family?
(A) Losaceae (B) Rosaceae
(C) Plumbagineceae (D) Compositae

1188. In Ikebana Shine is placed at the angel of ....
(A) 45 (B) 75
(C) 15 (D) 30

1189. Ikebana Shoe is placed at the angel of ....
(A) 45 (B) 75
(C) 15 (D) 30

1190. Ikebana Hikae is placed at the angel of ....
(A) 45 (B) 75
(C) 15 (D) 30

1191. Rose mutant Abhisarka is developed from ....
(A) Kiss of fire (B) Gulzar
(C) Banjaran (D) Arjun

1192. For cascade type bonsai which type container is suitable?
(A) Square (B) Oval
(C) Round (D) Rectangular

1193. What is the ideal PH for rose growing?
(A) 6-7.5 (B) 3-4.5
(C) 4-5 (D) 7.5-8.5

1194. Wild garden developed by William Robinson in .... century
(A) 18th (B) 19TH
(C) 20TH (D) 17TH

1195. Devil's tree is also known as.........
(A) Saptaprani (B) Amaltas
(C) Arjun (D) None of above

1196. *Alstonia scholaris* is botanical name of ....
(A) Arjun (B) Balam khira
(C) Devil's tree (D) Champa

1197. Mame bonsai height should not be more than .... cm
(A) 2-5 (B) 5-15
(C) 15-20 (D) 20-30

1198. How many flowers produced in double type Gerbera in ployhouse in per square metre?
(A) 400 (B) 250
(C) 300 (D) 200

1199. *Aracuria cookii* is highly suitable for ....
(A) Hedge (B) Edge
(C) Specimen (D) Screening

1200. Ascocenda is the hybrid of which flower?
(A) Gerbera (B) Orchids
(C) Gladiolus (D) Tuberose

1201. Which of the following is most efficient irrigation method for flower crops?
(A) Sprinkler (B) Drip
(C) Flood (D) Trench

1202. Dimorphotheca is very good for ....
(A) Hanging baskets (B) Bedding
(C) Pots (D) Bothe b & c

1203. Rose variety Mrinalini released from ....

(A) IARI (B) IIHR

(C) NBRI (D) PAU

1204. Which fungicide is used to prevent rose plant from dieback?

(A) Rose paint (B) Bavistin

(C) Captan (D) Zineb

1205. Which type Chrysanthemum is used for garlands and veni bracelets?

(A) Standard (B) Spray

(C) Pompon (D) Posy

1206. Central institute of medicinal and aromatic plants is located at ....

(A) Lucknow (B) New Delhi

(C) Gurgaon (D) Punjab

1207. *Alyssum maritimum* is an .... blooming annual

(A) Early (B) Late

(C) Mid (D) Very late

1208. Which type of gardening system was developed by Dr. Nathanial Ward?

(A) Hanging baskets (B) Vertical

(C) Terrariums (D) Jar and bottle gardening

1209. French Marigold is an ....

(A) Diploid (B) Tetraploid

(C) Triploid (D) Hexaploid

1210. Which colour flower bears in *Antigonon leptopus*?

(A) Red (B) Pink

(C) White (D) Blue

1211. Kamini and Shashank are the verities of ....

(A) Tuberose (B) China aster

(C) Marigold (D) Rose

1212. Leading flowers seed producing state in India?

(A) Haryana (B) Tamil Nadu

(C) West Bengal (D) Punjab

1213. Portulaca cultivar Jhumka is developed at?

(A) IIHR (B) NBRI

(C) PAU (D) IARI

1214. Yellow is contrasting to which colour?

(A) Orange (B) Violet

(C) Blue (D) Red

1215. Pseudo bulbs are found in which flower crop?
(A) Tuberose (B) Canna
(C) Orchid (D) All of these

1216. Gladiolus is propagated by?
(A) Bulb (B) Tuber
(C) Corm (D) Rhizome

1217. Swollen end of stolon is known as?
(A) Tuber (B) Bulbils
(C) Cormel (D) None of these

1218. Gladiolus cultivar Shobha is developed through which method?
(A) Hybridization (B) Selection
(C) Mutation (D) None of these

1219. Dahlia is propagated through?
(A) Tuber (B) Rhizome
(C) Cuttings (D) Tuberous root

1220. Swami vinyanand was related to which flower crop?
(A) Dahlia (B) Gladiolus
(C) Chrysanthemum (D) Hibiscus

1221. Scoring is important practice of which flower crop?
(A) Tuberose (B) Gladiolus
(C) Dahlia (D) All of these

1222. Planting time for spring flowering crop?
(A) January (B) December
(C) November (D) August

1223. Mayur variety of Gladiolus is released from?
(A) IIHR (B) NBRI
(C) IARI (D) AU Punjab

1224. Which of the following is national flower of Italy?
(A) Lilium (B) Tulip
(C) Lily (D) Tuberose

1225. How many chromosome numbers are there in Gladiolus?
(A) 30 (B) 40
(C) 60 (D) 16

1226. Negative geotropism in Gladiolus is occur during?
(A) Vegetative period (B) Storage
(C) Transporting (D) Flowering

1227. Breaking virus in tulip is transferred through?
(A) White fly (B) Thrips
(C) Aphids (D) Honey bee

1228. Which of the following is dicotyledonous bulbs plant?
(A) Gladiolus (B) Oxalis
(C) Orchids (D) All of these

1229. *Acroclinum roseam* is a botanical name of which flower?
(A) Dog flower (B) Duck flower
(C) Paper flower (D) None of these

1230. Which of the following is does not have modified organs to climb?
(A) Climbers (B) Crippers
(C) Twiners (D) Both B & C

1231. *Dracaena reflexa* is botanical name of which plant?
(A) Acalypha (B) Poinsettia
(C) Song of India (D) Dancing lady

1232. Rolling of lawn is done to?
(A) Levelling of ground (B) Touch nodes with ground
(C) Cut the grass (D) All of these

1233. Which of the following lawn grass does not require moving?
(A) Bermuda grass (B) Zoysia grass
(C) Agrostis grass (D) None of these

1234. Fairy ring spot in lawn grass is due to?
(A) Virus (B) Nutrient deficiency
(C) Fungus (D) Bacteria

1235. International registration authority of Rose is situated at?
(A) USA (B) China
(C) New Delhi (D) Bulgaria

1236. Which state produce highest cut flower in India?
(A) Karnataka (B) Tamil Nadu
(C) Rajasthan (D) West Bengal

1237. Hogarth course is known as?
(A) Line of purity (B) Line of beauty
(C) Line of trees (D) All of these

1238. Tulip garden is situated at?
(A) Srinagar (J & K) (B) Lucknow
(C) Agra (D) Lahore

1239. Which of the following flower is symbol of luxury?
(A) Carnation (B) Stock
(C) Statice (D) Lily

1240. Which of the following tree is related to Buddha?
(A) Ashoka (B) Sal
(C) Palash (D) All of these

1241. Pinjor garden Haryana is developed by?
(A) King Hayder Ali (B) Fadai Khan
(C) Bhuphindra Singh (D) Akbar

1242. Poet Kali Das mentioned which plant in play Shakuntala?
(A) Ashoka (B) Pipal
(C) Madhvi (D) Kadam tree

1243. Who was Le Notre?
(A) Architecture (B) English garden designer
(C) French garden designer (D) All of these

1244. Which of following annual is suitable for screening propose?
(A) Hollyhock (B) Sweet pea
(C) Phlox (D) Both A & B

1245. Which of following colour is dominant throughout the year in garden?
(A) Green (B) Red
(C) White (D) Grey

1246. Gomphrena belongs to which family?
(A) Umbeliferae (B) Amaranthaceae
(C) Rutaceae (D) Astraceae

1247. Which of the following is the cheapest method of lawn making?
(A) Dibbling of roots (B) Turfing
(C) Bricking (D) Seeding

1248. Which of the following flower is known as barbeton daisy?
(A) Gladiolus (B) Gerbera
(C) Tulip (D) Pansy

1249. Fluoride injury mostly found in which flower crop?
(A) Tuberose (B) Gladiolus
(C) Rose (D) All of these

1250. Tinting is an important practice of which flower crop?
(A) Rose (B) Carnation
(C) Hibiscus (D) Jasmine

1251. Which of the following bulbous crop cover highest area in India?
(A) Tuberose (B) Dahlia
(C) Gladiolus (D) Canna

1252. Which of the following is a thorn less variety of Rose?
(A) Suchitra (B) Grand Gala
(C) Pusa Mohit (D) All of these

1253. African Marigold is a symbol of?
(A) Vulgar mind (B) Jealousy and Sorrow
(C) Self-Steam (D) Purity

1254. Who was developed herbaceous border?
(A) G. Jekwell (B) W. Robinson
(C) BK. Roy (D) Le Notre

1255. Agave is used for which type of hedge?
(A) Dwarf ornamental (B) Tall ornamental
(C) Dwarf protective (D) Both A & B

1256. Mahua tree belong to which family?
(A) Acantheceae (B) Anacardiaceae
(C) Sapotaceae (D) Rutaceae

1257. Which colour flower found in *Bauhinia alba*?
(A) Purple (B) White
(C) Red (D) Yellow

1258. Which type of climbing organs present in Bougainvillea?
(A) Tendrils (B) Thorns
(C) Rootlets (D) None of these

1259. English flower arrangement is known as?
(A) Moribana (B) Morimona
(C) Jiyubana (D) Both A & B

1260. Optimum spacing for hybrid tea Rose?
(A) 60 cm (B) 90 cm
(C) 75 cm (D) 65 cm

1261. Which of the following is a thorn less species of Rose?
(A) *R. pendulina* (B) *R. bourboriana*
(C) *R. canina* (D) Both A & B

1262. Which of the following is a multi-coloured mutant variety of Rose?
(A) Abhisharka (B) Madhos
(C) Priya (D) All of these

1263. Hilling is an important operation of which flower crop?
(A) Tuberose (B) Chrysanthemum
(C) Carnation (D) Gladiolus

1264. Which of the following Carnation developed through cross between *Dianthus caryophyllus* X *Dianthus chinensis*
(A) Marguerite (B) Perpetual
(C) Royal (D) None of these

1265. Haldi Ghati is a variety of which flower?
(A) Rose (B) Dahlia
(C) Chrysanthemum (D) Carnation

1266. Which of the following flower is known as Rose of Indies?
(A) Hibiscus (B) China aster
(C) Marigold (D) Dahlia

1267. China aster is propagated through?
(A) Cuttings (B) Micro propagation
(C) Seeds (D) Rhizome

1268. Which of the following orchid is monopodial?
(A) Dendrobium (B) Vanda
(C) Cattleya (D) Bulbophyllum

1269. Which type of fruits are present in Rose?
(A) Capsule (B) Berry
(C) Nut (D) None of these

1270. Maximum oil recovery is done by which species?
(A) *J. sambac* (B) *J. Grandiflorum*
(C) *J. auriculatum* (D) *J. multiflorum*

1271. Mickey mouse is a variety of which flower?
(A) Carnation (B) Orchid
(C) Anthurium (D) Portulaca

1272. Chrysanthemum cultivar Money Maker is developed through?
(A) Biotechnology (B) Selection
(C) Hybridization (D) Mutation

1273. Which of the following first variety of Floribunda Rose?
(A) La France (B) Rodhatte
(C) Buccaner (D) Grand gala

1274. How much seeds of Marigold are required for one hectare?
(A) 1-1.5 Kg (B) 4-5 Kg
(C) 500 gm (D) 300 gm

1275. Which of the following type SI found in Marigold?
(A) Protandary (B) Protogyani
(C) Both A & B (D) None of these

1276. Which of the following largest genera of Orchid?
(A) Bulbophyllum (B) Dendrobium
(C) Vanda (D) Oncidium

1277. *Zoysia tennifolia* is a botanical of which lawn grass?
(A) Manilla grass (B) Korean grass
(C) Korean velvet grass (D) Carpet grass

1278. Which of the following storage disorder of Rose?
(A) Bull head (B) Bent neck
(C) Sleepiness (D) None of these

1279. Which flower is the largest flower of the world?
(A) Lotus (B) Arun Titan
(C) Gurhal (D) Delphinium

1280. Which of the following Rose variety have great demand in cut flower market?
(A) Raktagandha (B) Sindhoor
(C) Arjun (D) All of these

1281. Big sized bottles with narrow mouth are used for house plants is known as?
(A) Aquarium (B) Terrarium
(C) Pots (D) Canes

1282. Which of the following annual suitable for hanging baskets?
(A) Verbena (B) Petunia
(C) Zebrina (D) All of these

1283. Arrangement of potted colourful plants around the plant is known as?
(A) Topiary (B) Trophy
(C) Rhythm (D) Carpet bedding

1284. Who was introduced Mughal garden in India?
(A) Akbar (B) Babar
(C) Noorjhan (D) Hyder Ali

1285. In loamy soil Gerbera can be grown with ……….. Ph
(A) 7.5-8 (B) 6-6.5
(C) 4-5.5 (D) Above 8

1286. Axis is powerful in ………..
(A) English garden (B) Mughal garden
(C) Japanese garden (D) French garden

1287. Chrysanthemum is propagated through ....
(A) Terminal cuttings (B) Hard wood cutting
(C) Soft wood cutting (D) Budding

1288. Which of the following shrub is grown as an indoor pot plant?
(A) Acalypha (B) Beloperone
(C) Acacia (D) Euonymus

1289. In Rose Vit C is found in which part?
(A) Pods (B) Rose hips
(C) Leaves (D) Petals

1290. How much concentration of sugar is used for pulsing of Gladiolus?
(A) 10% (B) 20%
(C) 15% (D) 2%

1291. Beautiful gardens was written by ....
(A) B.P. Pal (B) M.S. Randhawa
(C) T.K. Boss (D) K.L. Chadda

1292. Which of the following is most suitable for ornamental dwarf hedge?
(A) *Acalypha sp.* (B) *C. tweediana*
(C) Agave (D) *A. farnesiana*

1293. Optimum isolation distance for cross pollination flower crops?
(A) 100m (B) 500m
(C) 1000m (D) 200m

1294. Arjuna is an mutant cultivar of ....
(A) Bougainvillea (B) Gladiolus
(C) Marigold (D) None of above

1295. Yellowing of lawn is due to ....
(A) Water logging (B) Fungal attack
(C) Virus attack (D) Insect attack

1296. What is the optimum planting distance for Polyanthus rose?
(A) 75 cm (B) 60 cm
(C) 45 cm (D) 30 cm

1297. What is the optimum planting distance for Hybrid tea rose?
(A) 75 cm (B) 60 cm
(C) 45 cm (D) 30 cm

1298. What is the optimum planting distance for Floribundas rose?
(A) 75 cm (B) 60 cm
(C) 45 cm (D) 30 cm

1299. For dormancy breaking Gladiolus corms should be stored at ....°C for 3-4 months

(A) 2-3 (B) 4-5

(C) 3 4 (D) 5-6

1300. Best time for Gladiolus planting in northern plains is ....

(A) Jan-Feb (B) Nov-Dec

(C) Sept-Oct (D) July-Aug

1301. Gladiolus hybrid 'Apsara' produce a spike length of .... cm

(A) 90 (B) 95

(C) 97 (D) 92

1302. Art of training plants into different shapes is known as ....

(A) Hedge (B) Edge

(C) Topiary (D) Trophy

1303. Fairy ring disease in lawn is caused by ....

(A) Virus (B) Fungus

(C) Insect (D) None of above

1304. Optimum seed rate of China aster is .... kg per hectare

(A) 1-1.5 (B) 2-2.5

(C) 2.5-3 (D) 3-3.5

1305. Begum Ali Yavar jung is a variety of ....

(A) Cooks' comb (B) Gladiolus

(C) Bougainvillea (D) Carnation

1306. Shin, Soe and Hikae are associated with ....

(A) Ikebana (B) Potpourri

(C) Pot culture (D) Bonsai

1307. Area of garden which is devoted exclusively for shrubs is known as ....

(A) Shrubbery (B) Hedge

(C) Edge (D) Carpet

1308. *Gynura aurantiaca* has .... colour foliage

(A) Yellow (B) Violet

(C) Golden (D) Pink

1309. Ayashii is the .... colour variety of Anthurium

(A) Red (B) Pink

(C) White (D) Yellow

1310. Avo-Claudia is the .... colour variety of Anthurium

(A) Red (B) Pink

(C) White (D) Yellow

1311. Avo-Jose is the .... colour variety of Anthurium
(A) Red (B) Pink
(C) White (D) Yellow

1312. Favoriet is the .... colour variety of Anthurium
(A) Orange (B) Pink
(C) White (D) Yellow

1313. Jamaica is the .... colour variety of Anthurium
(A) Orange (B) White
(C) Bright pink (D) Bright orange

1314. Jamaica is the .... colour variety of Anthurium
(A) Orange (B) White
(C) Bright pink (D) Bright orange

1315. Ozaki is the .... colour variety of Anthurium
(A) Red (B) Pink
(C) White (D) Yellow

1316. Manova-mist is the variety of ....
(A) Orchid (B) Gerbera
(C) Anthurium (D) Rose

1317. Top dressing of lawn is done in ....
(A) December (B) July
(C) February (D) August

1318. Pinching is required in ....
(A) Tuberose (B) Gladiolus
(C) Dahlia (D) Rose

1319. Optimum light intensity for Vanda Orchid is .... foot candle
(A) 800 (B) 500
(C) 1000 (D) 1500

1320. Ideal temperature range for indoor plants is .... °C
(A) 18-25 (B) 15-21
(C) 10-12 (D) 25-30

1321. The Oblique type bonsai is also known as ....
(A) Chokkan (B) Hankan
(C) Kengai (D) Kyokkok

1322. Formal upright type bonsai is also known as ....
(A) Chokkan (B) Hankan
(C) Kengai (D) Kyokkok

1323. In formal upright type bonsai is also known as ....

(A) Chokkan (B) Hankan

(C) Moyogi (D) Kyokkok

1324. Double trunk type bonsai is also known as ....

(A) Chokkan (B) Sokan

(C) Moyogi (D) Kyokkok

1325. Multiple trunk type bonsai is also known as ....

(A) Chokkan (B) Sokan

(C) Moyogi (D) Kabudachi

1326. Forest type bonsai is also known as ....

(A) Yose ue (B) Sokan

(C) Moyogi (D) Kabudachi

1327. Raft type bonsai is also known as ....

(A) Chokkan (B) Ikadabuki

(C) Moyogi (D) Kabudachi

1328. Broom type bonsai is also known as ....

(A) Chokkan (B) Ikadabuki

(C) Hokidachi (D) Kabudachi

1329. Cascade type bonsai is also known as..........

(A) Kengai (B) Sokan

(C) Moyogi (D) Kyokkok

1330. Semi-Cascade type bonsai is also known as ....

(A) Kengai (B) Han-Kengai

(C) Moyogi (D) Kyokkok

1331. Windswept type bonsai is also known as ....

(A) Fukingashi (B) Han-Kengai

(C) Moyogi (D) Kyokkok

1332. Bonsai in rock is also known as ....

(A) Seki joju (B) Ishusuki

(C) Moyogi (D) Kyokkok

1333. Which of the following is largest exporter of cut flowers?

(A) Germany (B) India

(C) Netherland (D) Japan

1334. First plant cloned was developed in ....

(A) Carnation (B) Orchid

(C) Gerbera (D) Gladiolus

1335. Which climber is suitable for trellis for the purpose only as a light climber?
(A) *Allamanda cathartica* (B) *Adenocalymma allicea*
(C) *Petrea volubilis* (D) *Solanum seaforthianum*

1336. Which of the following climber has foliage habitat?
(A) *Monstera deliciosa* (B) *Passiflora edulis*
(C) *Jasminum. Sp.* (D) *Jacquemontia violaceae*

1337. Which of the following is suitable for grow as a edge purpose?
(A) Eupatorium (B) Alternenthera
(C) Duranta (D) All the above

1338. The plant which is suitable for Topiary purpose?
(A) Cupressus (B) Tecoma
(C) Graptophyllum (D) Alyseum

1339. Which tree is suitable for roof gardening?
(A) Canna (B) Anthirrhinum
(C) Mimusops (D) Vernonia

1340. Plant suitable for water gardening
(A) Nuphar luteum (B) Nymphaea spp.
(C) Duck weed (D) All the above

1341. Which plant is suitable for Bog gardens?
(A) Sagittaria (B) Cupressus
(C) Alternenthera (D) Pelargonium

1342. The plant suitable for shade garden
(A) Digitalis (B) Candytuft
(C) Ageratum (D) Magnolia

1343. Plant suitable for dish garden
(A) Snake plant (B) Azalea
(C) Salix (D) Camellia

1344. Which one is white fragrant flower tree?
(A) *Casuarina equisetifolia* (B) *Grevilea robusta*
(C) *Millingtonia hortensis* (D) *Butea monosperma*

1345. Which of the following is a winter annual?
(A) Balsam (B) Petunia
(C) Kochia (D) Tithonia

1346. Which of the following is Cycad?
(A) *Caryota urens* (B) Zamia
(C) Pritchardia (D) Adiantum

1347. The plant part which is modify into a spine
(A) Flowers (B) Roots
(C) Leaves (D) Stem

1348. Rock plants need replanting in .... years
(A) 4-5 (B) 2-3
(C) 3-4 (D) 5-6

1349. Polypodium is an ....
(A) Succulent (B) Cacti
(C) Fern (D) Winter annual

1350. Agave belong to the family ....
(A) Euphorbiaceae (B) Aspragusceae
(C) Amaryllidaceae (D) Amranthaceae

1351. Suitable shed net for Codiaeum
(A) White (B) Red and Grey
(C) Green (D) All of above

1352. *Alpinio speciose* is an ....
(A) Cacti (B) Succulent
(C) Bulbous plant (D) Fern

1353. *Beaumontia grandiflora* produce .... colour flowers
(A) Yellow (B) White
(C) Red (D) Pink

1354. In which type of Bonsai have single trunk and twisted once or more times?
(A) Cascade (B) Oblique
(C) Upright (D) Winding

1355. M.A. Kher was associated with ....
(A) Gladiolus (B) Chrysanthemum
(C) Marigold (D) Dahlia

1356. Growing of Flowers in standing acid water is known as ....
(A) Bog garden (B) Rockery garden
(C) Flat garden (D) None of these

1357. Which of the following Ornamental plant not native to India?
(A) Hedychum (B) Lonicera
(C) Jacaranda (D) Tabernaemontana

1358. Miracle garden is located at.........
(A) London (B) Dubai
(C) Bagdad (D) England

1359. Which of the following country have highest per capita consumption of cut flower?
(A) Switzerland (B) Netherland
(C) Germany (D) China

1360. Who was the first Mughal empire introduced Mughal garden in India?
(A) Akbar (B) Babar
(C) Jahangir (D) Sahanjha

1361. Following cultivar of Bougainvillea not developed through interspecific hybridization?
(A) Chitra (B) Wazi Ali Shah
(C) Matama Gandhi (D) Begaum Sikandar

1362. In which type of garden the element of surprise is lost?
(A) Informal (B) Formal
(C) Persian (D) Wild garden

1363. Which of the following is not a symmetrical garden?
(A) Rashtrapati Bhavan (B) Pinjor garden
(C) Taj Mahal (D) Budh Jayanti park

1364. Which of the following is a deciduous tree?
(A) Juniperus (B) Criptomeria
(C) Salix (D) Magnolia

1365. Which of the following is cabbage Rose?
(A) R. foitida (B) R. gallica
(C) R. centifolia (D) R. canina

1366. High quality perfume is produced from............
(A) J. sambac (B) J. multiflora
(C) J. grandiflorum (D) J. auriculatum

1367. Sweet William is belong to family ....
(A) Rutaceae (B) Bracicaceae
(C) Caryophyllaceae (D) Astraceae

1368. Stock is belong to family ....
(A) Bracicaceae (B) Araceae
(C) Plumbaginaceae (D) Nactaginaceae

1369. Arka Ambra is a variety of ....
(A) Gladiolus (B) Crossandra
(C) Carnation (D) Dahlia

1370. Pusa Basanti is a variety of ....

(A) Chrysanthemum (B) Marigold

(C) Dahlia (D) Petunia

1371. Parrot flower is also known as ....

(A) Heliconia (B) Iris

(C) Sweet sultan (D) Sweet William

1372. How much Rose petals required for 1 Kg Rose oil?

(A) 10 Ton (B) 3-4 Ton

(C) 10-15 Ton (D) 8-10 Ton

1373. Precursor of Ethylene ....

(A) Methionine (B) Tetracycline

(C) Terthelion (D) None of these

1374. Which of the following is a variety of Tuberose?

(A) PG purple (B) PG White

(C) Arka Nirantra (D) Arka Kamini

1375. Which of the following is not the principle of landscaping?

(A) Line (B) Simplicity

(C) Balance (D) Accent

1376. Informal edging is made by ....

(A) Tiles (B) Bricks

(C) Plants (D) All of above

1377. Optimum room temperature for pulsing is .... ºC

(A) 25-30 (B) 20-27

(C) 15-20 (D) 10-12

1378. Rose cultivar Jwala is released from ....

(A) IIHR (B) IHBT

(C) IARI (D) NBRI

1379. Rose cultivar Himroz is released from ....

(A) IIHR (B) IHBT

(C) IARI (D) NBRI

1380. Highest ascorbic acid found in which species of rose?

(A) *R. gallica* (B) *R. rugose*

(C) *R. Centifolia* (D) *R. moschata*

1381. Which of the following is scented foliage species of rose?

(A) *R. gallica* (B) *R. rubiginosa*

(C) *R. Centifolia* (D) *R. moschata*

1382. Queen Elizabeth variety of rose related to ....
(A) Hybrid tea (B) Grandiflora
(C) Floribundas (D) Polyanthus

1383. Chrysanthemum was introduced by Britishers in .... Century
(A) 17TH (B) 18TH
(C) 19TH (D) None of above

1384. Neelima variety of Chrysanthemum released from ....
(A) IARI (B) IIHR
(C) NBRI (D) DFR

1385. Crown bud's production in Chrysanthemum is due to ....
(A) Longer day length (B) Short day length
(C) Insect attack (D) Fungal attack

1386. Chrysanthemum is an .... plant
(A) Quantitative short day (B) Quantitative long day
(C) Qualitative short day (D) Qualitative long day

1387. Carnation is an .... Plant
(A) Quantitative short day (B) Quantitative long day
(C) Qualitative short day (D) Qualitative long day

1388. Arka Tejas variety of carnation has been developed through ....
(A) Interspecific hybrid (B) Intraspecific hybrid
(C) Intergeneric hybrid (D) Mutation

1389. Buds don't open evenly in carnation is the symptoms of........
(A) Calyx splitting (B) Slabside
(C) Curly tip (D) None of these

1390. How many genera are recorded in Orchidaceae family?
(A) 500 (B) 700
(C) 750 (D) 800

1391. Which of the following is second largest genus of Orchidaceae family?
(A) Dendrobium (B) Bulbophyllum
(C) Oncidium (D) Vanda

1392. *Oncidium* orchid is also known as ....
(A) Dancing lady (B) Moth orchid
(C) Lady slipper (D) Fox tail orchid

1393. *Paphiopedilum* orchid is also known as ....
(A) Dancing lady (B) Moth orchid
(C) Lady slipper (D) Fox tail orchid

1394. *Rhyncostylus* orchid is also known as ....
(A) Dancing lady (B) Moth orchid
(C) Lady slipper (D) Fox tail orchid

1395. Gerbera is originated from ....
(A) South Africa (B) Mexico
(C) China (D) East Asia

1396. Gladiolus is also known as ....
(A) Sword lily (B) Corn flag
(C) Both a & b (D) Crown daisy

1397. Which from the following options is NOT a result of Aphid in a plant?
(A) Premature leaf fall (B) Vigour loss
(C) Flower damage (D) Yellowing

1398. When is the flower blooming time of *Jasminum multiflorum*?
(A) Nov-Feb (B) May-June
(C) March-April (D) None of above

1399. Which symbolism of Moghul Garden has been taken from Koran/Quran?
(A) Six division (B) Eight division
(C) Seven division (D) Nine division

1400. What should be added in soil from the given options when it is very acidic?
(A) Lime (B) Urea
(C) Salt (D) DAP

1401. Optimum potassium nitrate requirement for Asiatic Liliums in g/m^2/week?
(A) 2.2 (B) 1.5
(C) 1.8 (D) 3.5

1402. What quantity of NPK required for Chrysanthemum propagation per hectare?
(A) 20 kg (B) 25 kg
(C) 30 kg (D) 35 kg

1403. The seeds of *Ammi Majus* start to germinate within ________ days after sowing.
(A) 10-12 (B) 4-5
(C) 7-10 (D) 15-20

1404. What is the optimum day length required for Gladiolus cultivation?
(A) 12-14 hours (B) 8-10 hours
(C) 15-16 hours (D) Any of above

1405. Growing of trees both side of roads is called as ....
(A) Topiary (B) Trophy
(C) Avenue (D) Archis

1406. **Which of the following group is not related to annual plant classification?**

(A) Winter season (B) Dry weather

(C) Summer season (D) Rainy season

1407. **Which of the following temperature range is right for Tuberose cultivation?**

(A) 28-30 ºC (B) 20-25 ºC

(C) 30-32 ºC (D) 35-40 ºC

1408. **Which of the following is the correct NPK doze for Gladiolus?**

(A) 60:150:150 (B) 80:30:50

(D) 30:100:110 (D) 50:120:120

1409. **What should be the ideal soil pH for Chrysanthemum to grow?**

(A) 1.5-2 (B) 6.5-7

(C) 4.2-5.5 (D) 7.5-8.5

1410. **Required NPK ration for Tuberose cultivation is …………**

(A) 200:200:200 (B) 150:150:150

(C) 120:100:200 (D) 80:150:150

1411. **Which of the following is not used to grow Ipecac?**

(A) Tuber (B) Seeds

(C) Roots (D) Leaf cutting

1412. **Dahlia is a half hardy ……….. plant**

(A) Annual (B) Bi-annual

(C) Perennial (D) Monthly

1413. **To control lesion Nematode what is the required doze of Carbofuran?**

(A) 25 kg/ha (B) 20 kg/ha

(C) 22 kg/ha (D) 30kg/ha

1414. **Optimum required day temperature for Lilium?**

(A) 15-18 ºC (B) 18-22 ºC

(C) 20-25 ºC (D) 25-28 ºC

1415. **Growing media Perlite has ………. moisture holding capacity**

(A) High (B) Low

(C) Medium (D) Very low

1416. **Gladiolus is commercially propagated through ……….**

(A) Corms (B) Tuber

(C) Rhizome (D) Bulb

1417. **Which of the following plants is NOT suitable for water gardens?**

(A) *Victoria regia* (B) *Nerium oleander*

(C) Lemnagibbba (D) *Elodea canadensis*

1418. How much quantity of Copper Oxychloride mixed with 1 litre water to control root rot?
(A) 1.5 g/lit (B) 2.5 g/lit
(C) 3.5 g/lit (D) 3 g/lit

1419. Chlorosis of plants is caused due to deficiency of ....
(A) Magnesium and Iron (B) Boron and Potassium
(C) Chorine and copper (D) Carbon and Nitrogen

1420. What should be the optimum size of pits for Rose?
(A) 2x2 m (B) 2x1 m
(C) 2x3 m (D) 2x4 m

1421. Viola species have __________ leaves and flowers.
(A) Round and small (B) Large
(C) mild-flavoured edible (D) Poisonous

1422. Which of the following is not the example of Asexual propagation?
(A) Seeds (B) Cuttings
(C) Roots (D) Tuber

1423. What should be the ideal length of tip for the propagation of Rose?
(A) 6 to 8 inch (B) 2 to 5 inch
(C) 8 to 10 inch (D) 1 to 2 inch

1424. Which part of tuberose is used for commercial propagation?
(A) Bulb (B) Tuber
(C) Rhizome (D) Corm

1425. Which of the following is NOT a type of Rajavanam?
(A) Promododyan (B) Vriksha Vatika
(C) Brindavanam (D) Udayan

1426. Evergreen plant for hedge should be planted at .... feet distance
(A) 1-2.5 (B) 2-3
(C) 2-3.5 (D) 3-4

1427. In dry areas sunken bed should be prepared .... cm below ground level
(A) 8-10 (B) 10-15
(C) 7-12 (D) 15-20

1428. What is the crop duration of Oriental hybrid Lilium?
(A) 10-12 Weeks (B) 14-16 Weeks
(C) 8-10 Weeks (D) 15-20 Weeks

1429. According to ICAR Bluegrass best adapted a soil PH between ....
(A) 6-7 (B) 4-5
(C) 4.5-6.5 (D) 7-8

1430. What is the ideal pH of soil for growing roses?

(A) 6-7 (B) 4-5
(C) 4.5-6.5 (D) 7-8

1431. To cultivate tuberose, well drained loamy soil having a pH of_________ is ideal.

(A) 6.5-7.5 (B) 4-5
(C) 4.5-6.5 (D) 7-8

1432. Lilium bulbs should be stored at 2 to 5 ºC for .... Weeks for dormancy breaking

(A) 5 (B) 6
(C) 4 (D) 1

1433. What should be the pH of the soil to grow Gladiolus?

(A) 6.5-7.0 (B) 4-5
(C) 4.5-6.5 (D) 7-8

1434. A special dried flower arrangement is known as..........

(A) The Crescent shape model (B) Potpourri
(C) The Oval bowl model (D) Flower bouquet

1435. Chrysanthemum is commercially propagated by ....

(A) Leaf cutting (B) Bulb
(C) Terminal cutting (D) Seeds

1436. To control leaf spot in Gladiolus, _________ Mancozeb is added per litre and sprayed.

(A) 1 gm (B) 2 gm
(C) 3 gm (D) 4 gm

1437. Dahlia is native of which country?

(A) Russia (B) Mexico
(C) China (D) South Africa

1438. Which of the following is not a Bonsai art?

(A) Full cascade (B) Formal upright
(C) Rock cascade (D) Wind swept

1439. What should be angle at which rose should be cut for root?

(A) 35º (B) 45º
(C) 60º (D) 75º

1440. Cool season grasses grow at a temperature between .... ºC

(A) 10-25 (B) 40-50
(C) 15-30 (D) 25-35

1441. Which of the following is a type of an informal garden?
(A) Mughal (B) Japanese
(C) French (D) Persian

1442. What should be the size of terminal cutting for propagation of Chrysanthemum?
(A) 2-3 cm (B) 5-7 cm
(C) 4-5 cm (D) 8-10 cm

1443. Showing time of Crossandra seeds is ....
(A) June-July (B) Oct-Nov
(C) Dec-Jan (D) March-April

1444. Which of the following CANNOT be used as a ground cover plant?
(A) Rock cress (B) Mulberry
(C) Carpet bugle (D) Common Yarrow

1445. White Kachnar represent the which things?
(A) Extra Ordinary (B) Youth & Life
(C) Self-respect (D) All of these

1446. Rashtrapti Bhavan garden is which type of garden?
(A) Japanese (B) English
(C) Mughal (D) None of these

1447. Which of the following is not a type of Japanese garden?
(A) Hill garden (B) Flat garden
(C) Tea garden (D) Forcing garden

1448. Who was coined term Bio aesthetic planning?
(A) Prof. Lancelot Hogben (B) W. Robinson
(C) Le Notre (D) None of these

1449. Optimum size for Bonsai?
(A) 20-30 cm (B) 30-50 cm
(C) 30-60 cm (D) 50-60 cm

1450. Floral clock is a special feature of which garden?
(A) Pinjor garden (B) Rock garden
(C) Lal Bagh garden (D) All of these

1451. Optimum drying temperature for Chrysanthemum?
(A) 45-49°C for 72 hours (B) 45-49°C for 48 hours
(C) 45-49°C for 96 hours (D) 45-49°C for 25 hours

1452. Which of the following is an off season variety of Tuberose?
(A) Rajat Rekha (B) Shringar
(C) Suvashini (D) Pearl Double

1453. Nugget variety of Marigold was developed from?
(A) Egypt (B) USA
(C) India (D) Netherland

1454. Rose variety Mohini in India introduce from?
(A) USA (B) China
(C) Germany (D) Italy

1455. Inflorescence type of Chrysanthemum?
(A) Spike (B) Spadix
(C) Capitulum (D) None of these

1456. What is mean of pot mums?
(A) 1 Plant in 1 Pot (B) 1 Species in 1 Pot
(C) 1 Species in >1 Pot (D) 1 Plant in >1 Pot

1457. Who was introduce Damask Rose in India?
(A) Baber (B) British
(C) Araya (D) B.P. Pal

1458. Scoring is an important operation of which flower/
(A) Tulip (B) Gladiolus
(C) Hibiscus (D) Rose

1459. Hira-Niwa is the Japanese mean of which type Japanese garden?
(A) Sand garden (B) Flat garden
(C) Tea garden (D) Hill garden

1460. Which of the following plant is suitable for Bonsai?
(A) *Ficus religiosa* (B) *F. benjamina*
(C) *F. bengalensis* (D) All of these

1461. Which of the following is not a feature of English garden?
(A) Lawn (B) Running water
(C) Rockery (D) Herbaceous garden

1462. Buddha Jayanti park is a which type of garden?
(A) Mughal (B) Japanese
(C) English (D) Persian

1463. Which of the following is a best rooting media for Rose?
(A) Sand (B) Sand + FYM
(C) FYM + Coir (D) All of these

1464. Foliage Nematode is a serious problem of which flower?
(A) Tuberose (B) Gladiolus
(C) Orchid (D) All of these

1465. Which of the following type pollination found in Orchid?
(A) Animophilous (B) Hydrophilous
(C) Entomophilous (D) Ornithophilous

1466. Economic part of Anthurium?
(A) Leaves (B) Spathe
(C) Whole plant (D) Spadix

1467. Which of the following flower is an indicator of fluoride?
(A) Chandni (B) Gladiolus
(C) Tuberose (D) Kamini

1468. Disbudding in Dahlia is done at which stage?
(A) Bud appearance stage (B) Pea stage
(C) Bud opening stage (D) Flower opening stage

1469. Stem break is a post-harvest disorder of which flower?
(A) Dahlia (B) Gerbera
(C) Tuberose (D) Gladiolus

1470. Lotus is propagated through which method?
(A) Suckers (B) Seeds
(C) Cuttings (D) All of these

1471. Dr. B.P. Pal is a variety of which flower?
(A) Bougainvillea (B) Rose
(C) Chrysanthemum (D) Gerbera

1472. Which type of pollination found in Heliconia?
(A) Ornithophilous (B) Self pollination
(C) Entomophilous (D) Animophilous

1473. Which type of cross pollination mechanism found in Primula?
(A) Self Incompatibility (B) Dicogamy
(C) Heterostyle (D) Male sterility

1474. Which of the following annuals is/are Indian origin?
(A) Gomphrena (B) Balsam
(C) Lady's lack (D) All of these

1475. *Portulaca grandiflora* is a botanical name of which annual?
(A) Sun plant (B) Portulaca
(C) Sun flower (D) Both A & B

1476. Which of the following Rose is mainly used for gulkhand?
(A) Damask Rose (B) Chinese Rose
(C) Edward Rose (D) Dog Rose

1477. Colour break in Anthurium spathe is due to deficiency of?

(A) Calcium (B) Boron

(C) Nitrogen (D) Phosphorus

1478. Which of the following pollination agent in Petunia?

(A) Honey bees (B) Air

(C) Water (D) Birds

1479. Propagation in Anthurium is done through?

(A) Rhizome (B) Slips

(C) Suckers (D) Bulbs

1480. Yellow colour is extracted from which Rose species?

(A) *R. foetida* (B) *R. canina*

(C) *R. galica* (D) All of these

1481. Semal tree is related to?

(A) Shiva (B) Kali

(C) Krishna (D) Buddha

1482. Kamini is belongs to which family?

(A) Sapotaceae (B) Rutaceae

(C) Protceae (D) Oleaceae

1483. Shri Bajrang Bahadur Singh is related to which flower?

(A) Gladiolus (B) Chrysanthemum

(C) Jasmine (D) Gerbera

1484. Which of the following plant is/are suitable for indoor condition?

(A) Acalypha (B) Spider plant

(C) Poinsettia (D) All of these

1485. Which of the following is not a feature of informal garden?

(A) Land is not levelled (B) Plan is forced to fit the land

(C) Symmetrical design (D) Individual plants are not selected as feature

1486. Roses may be rooted at any time of the year, but success is much more likely during

(A) Nov to Feb (B) July to Aug

(C) March to April (D) May to Aug

1487. In case of transplanting a perennial which was grown in a pot, the root balls should be

(A) Washed clean (B) Left as it is

(C) Slightly turn (D) Cut off

1488. What is the ideal temperature at day to grow Chrysanthemum?

(A) 30-32°C (B) 20-28°C

(C) 20-25°C (D) 15-20°C

1489. Arka Krishika is a variety of ....

(A) Gladiolus (B) China aster

(C) Gerbera (D) Marigold

1490. Best time for Gerbera planting is ....

(A) May-July (B) Oct-Nov

(C) Dec-Jan (D) Any of above

1491. Extreme yellowing of young leaves of Gerbera is due to .... deficiency

(A) Boron (B) Calcium

(C) Zinc (D) Iron

1492. Gladiolus variety White Friendship is developed through ....

(A) Hybridization (B) Mutation

(C) Tissue culture (D) None of these

1493. Gladiolus variety Pusa Srijana was released from ....

(A) IIHR (B) PAU

(C) IARI (D) DFR

1494. Apsara variety of gladiolus was released from ....

(A) IIHR (B) IHBT

(C) CCSHAU (D) IARI

1495. Gazel is a variety of ....

(A) Gerbera (B) Gladiolus

(C) Rose (D) Tuberose

1496. Gladiolus variety Solan Mangla is developed through ....

(A) Mutation (B) Hybridization

(C) Introduction (D) Biotechnology

1497. Tuberose is originated from ....

(A) South Africa (B) Mexico

(C) China (D) Austria

1498. Tuberose variety Suvasini is released from ....

(A) IARI (B) BCKV, West Bengal

(C) IIHR (D) DFR

1499. Tuberose variety Rajat Rekha is developed through ....

(A) Mutation (B) Hybridization

(C) Introduction (D) Tissue culture

1500. Tuberose variety Rajat Rekha is released from ....
(A) NBRI (B) IIHR
(C) IARI (D) BCKV, West Bengal

1501. Which of the following variety of Tuberose has been released from BCKV, West Bengal?
(A) Bidhan Rajni-1 (B) Bidhan rajni-2
(C) Bidhan rajni-3 (D) All of above

1502. Tuberose cultivar GKTC-4 has been released from ....
(A) MPKV, Rahuri (B) BCKV, West Bengal
(C) IIHR, Bangalore (D) IARI, New Delhi

1503. Which of the following is result of Mexican single X Pearl Double?
(A) Arka Shringar (B) Arka Sugandhi
(C) Arka Nirantara (D) Arka Prajwal

1504. Showing time of tuberose in hilly areas is ....
(A) April-May (B) June-July
(C) Oct-Nov (D) Dec-Jan

1505. For dormancy breaking of Tuberose bulbs .... Thiourea is suitable
(A) 4% (B) 5%
(C) 6% (D) 8%

1506. .... Quintals of Tuberose bulbs required for one hectare
(A) 10-12 (B) 15-20
(C) 20-22 (D) 5-7

1507. Anthurium is originated from ....
(A) Columbia (B) China
(C) India (D) Mexico

1508. Suckers production will be start in Anthurium when they attain an age of........
(A) 12-16 Months (B) 10-12 Months
(C) 8-10 Months (D) 15-20 Months

1509. Anthurium can be easily stored at .... ºC
(A) 13 (B) 10
(C) 3-4 (D) 1-2

1510. Leaf scorching in Lilium is due to...........
(A) High temperature (B) Low temperature
(C) Temperature fluctuations (D) High humidity

1511. Lilium bulbs planted during ....
(A) Oct-Nov (B) Dec-Jan
(C) Feb-March (D) July-Aug

1512. Marigold is also known as ....
(A) Devine flower (B) Gilly flower
(C) Friendship flower (D) All of above

1513. Highest oil recovery species of Marigold is ....
(A) *T. lucida* (B) *T. minuta*
(C) *T. patula* (D) *T. erecta*

1514. Which of the following is a sweet scented species of Marigold?
(A) *T. lucida* (B) *T. minuta*
(C) *T. patula* (D) *T. erecta*

1515. Marigold variety Arka Agni has been released from ....
(A) IARI (B) IIHR
(C) NBRI (D) PAU

1516. Which of the following jasmine species is substitute for Saffron?
(A) *J. grandiflorum* (B) *J. multiflorum*
(C) *J. sambac* (D) *J. humile*

1517. Which of the following species of Jasmine especially suitable for home garden?
(A) *J. grandiflorum* (B) *J. multiflorum*
(C) *J. auriculatum* (D) *J. humile*

1518. Nearly 50 % of world jasmine oil produce from ....
(A) Bulgaria (B) China
(C) France (D) Egypt

1519. Crossandra is also known as ....
(A) Devine flower (B) Fig Marigold
(C) Fire cracker (D) Cape Marigold

1520. A.P.J. Abdul Kalam is a mutant variety of ....
(A) Carnation (B) Marigold
(C) Crossandra (D) Stock

1521. National flower of India is ....
(A) Blue Lotus (B) Pink Lotus
(C) Red Lotus (D) Yellow Lotus

1522. National flower of Mexico is ....
(A) Dahlia (B) Lily
(C) Statice (D) Pansy

1523. Dahlia cuttings are raised in ....
(A) Sept-Oct (B) Dec-Jan
(C) March-May (D) July-Aug

1524. Dahlia is a ………….. day plant

(A) Short (B) Long
(C) Short-Long (D) Day neutral

1525. The headquarters of National Horticulture Board is located at ………..

(A) Bengaluru (B) Gurgaon
(C) New Delhi (D) Pune

1526. Which of the following lawn grass species grows well in acids soil?

(A) Kikuyu grass (B) Fescue grass
(C) Manila grass (D) Dwarf Bermuda

1527. Which of the following is unusual chocolate coloured rose variety?

(A) Mohini (B) Rupali
(C) Chandrma (D) Neelambari

1528. The colour of december poovu is …………

(A) Yellow (B) Purplish blue
(C) Pink (D) White

1529. Bougainvillea cultivar Spring festival has been released from ………

(A) IARI (B) IIHR
(C) NBRI (D) FRI

1530. World's largest flower auction trade centre is at …………

(A) Alsmeer (B) Bengaluru
(C) Chandigarh (D) New Delhi

1531. As per September 2020 which state have highest area under floriculture?

(A) West Bengal (B) Karnataka
(C) Andhra Pradesh (D) Rajasthan

1532. Which of the following is purpose of growing *Grevillea robusta*?

(A) Aromatic (B) Medicinal
(C) Avenue (D) Edge

1533. Which of the following is not a type of Japanese garden?

(A) Flat garden (B) Tea garden
(C) Tomb garden (D) Hill garden

1534. Fragrant flowers of maramalli, which open in the evening, are ………. in colour

(A) Blue (B) White
(C) Yellow (D) Pink

1535. Optimum storage temperature for gladiolus corm is …….. °C

(A) 0 – 1 °C (B) 2 – 5 °C
(C) 4 – 7 ° (D) 10 -12 °C

1536. Which of the following is not a fragrant variety of Gladiolus?
(A) Lucky star (B) Jimmy boy
(C) Sager (D) All of these

1537. Punjab gold is a variety of ....
(A) Marigold (B) Chrysanthemum
(C) Jasmine (D) Gladiolus

1538. Storage corm rot in Gladiolus is caused by.........
(A) Fusarium (B) Pythium
(C) Panicillium (D) Rhizoctinia

1539. Arthur Sim Carnation is highly resistance to ....
(A) Wilt (B) Rust
(C) Stem rot (D) All of these

1540. Who was started tissue culture in Orchid?
(A) Lysenko (B) Shull
(C) Morel (D) S. Srivastava

1541. Calyx splitting in Carnation is due to ....
(A) Genetic factor (B) Environmental factor
(C) Nutritional factor (D) All of these

1542. Most serious pest of Carnation.......
(A) Aphids (B) Thrips
(C) Red spider mite (D) White fly

1543. Gulkhand is prepared by using petals and sugar in ratio of .....
(A) 1:1 (B) 1:2
(C) 2:1 (D) 1:3

1544. Chemical defoliation in Rose is done by ....
(A) Urea (B) $CuSo_4$
(C) GA3 (D) ABA

1545. Which colour Roses are most popular?
(A) Red (B) Pink
(C) Yellow (D) Blue

1546. In cut Roses minimum vase life should be .... days
(A) 10 (B) 12
(C) 8 (D) 14

1547. Tuberose cultivar Suvasini is released from ....
(A) IARI (B) NBRI
(C) IIHR (D) BCKV

1548. Dahlia tuberous roots should be stored in ....
(A) Cold storage (B) Sand
(C) Coir dust (D) CAS

1549. Spotted wilt in Dahlia is caused by ....
(A) Fungus (B) Virus
(C) Bactéria (D) Genetic

1550. Dr. B.P. Pal was developed how many cultivars of Rose?
(A) 100 (B) 102
(C) 200 (D) 105

1551. Normal humidity in green house should be .... percent
(A) 30 – 40 (B) 20
(C) 60 – 65 (D) 70 - 80

1552. Diploid Chromosome number in Gerbera........
(A) 30 (B) 50
(C) 60 (D) 18

1553. Anthurium flower can be stored at .... temperature
(A) 2 ºC (B) 13 ºC
(C) 5 ºC (D) 10 ºC

1554. Which is a serious disease of Anthurium?
(A) Rust (B) Wilt
(C) Anthracnose (D) All of these

1555. *Hemerocallis fulva* is botanical name of ....
(A) Day Lily (B) Sword Lily
(C) Tiger Lily (D) All of these

1556. Bird of Paradise is native to...........
(A) India (B) America
(C) South Africa (D) China

1557. China aster is a .... crop
(A) Self-pollinated (B) Cross pollinated
(D) Often cross pollinated (D) None of these

1558. Narcissus flowers are harvested at which stage?
(A) Tight bud stage (B) Goose neck stage
(C) Half open stage (D) Fully open stage

1559. Pin and Thrum type flowers are found in which flower?
(A) Petunia (B) Primula
(C) Poinsettia (D) Orchid

1560. Which one is fully compatible?

(A) $S_1 \times S_2 \times s_3 \times s_4$ (B) $S_1 \times S_2 \times S_3 \times S_4$

(C) $S_1 \times s_2 \times s_3 \times s_4$ (D) All of these

1561. Which of the following male sterility is commonly found in Ornamental plants?

(A) CGMS (B) CMS

(C) GMS (D) All of these

1562. Single gene male sterility is transferred in .... years

(A) 2–3 (B) 7–8

(C) 4-5 (D) 10

1563. Virus free plants produced through...........

(A) Anther culture (B) Meristem culture

(C) Seed (D) Cuttings

1564. Which aneuploids is mostly used in Ornamentals?

(A) Nullisomic (B) Monosomic

(C) Trisomic (D) Tetrasomic

1565. Nullisomic means ....

(A) 2n + 1 (B) 2n + 2

(C) 2n – 1 (D) 2n – 2

1566. Lilium bulbs are stored at............... ºC

(A) 2-3 (B) -2-2

(C) 8-10 (D) 7-8

1567. Representation of the structure of any surface is known as ....

(A) Shape (B) Canopy

(C) Texture (D) Scale

1568. Optimum water holding capacity of potting media should be .... percent

(A) 30–40 (B) 40-50

(C) 50-60 (D) 70-80

1569. Soluble salt in potting media should be .... ppm

(A) <100 (B) >100

(C) <500 (D) >500

1570. Which pots are most suitable for Cacti?

(A) Cement (B) Plastic

(C) Soil (D) All of these

1571. In normal air how much Ethylene is present?

(A) 5-10 ppb (B) 3-5 ppb

(C) 3-5 ppm (D) 100-200 ppb

1572. Bent neck in Rose is very common when grow at ....
(A) High light (B) Low light
(C) High temperature (D) Low temperature

1573. HQC is an effective ....
(A) Germicide (B) Fungicide
(C) Bactericide (D) Weedicide

1574. Salt of HQ is a most common ....
(A) Germicide (B) Fungicide
(C) Bactericide (D) Weedicide

1575. Green Glo is used for improve longevity of.........
(A) Flower (B) Foliage
(C) Fruit (D) All of these

1576. CO_2 concentration in normal air is about...........
(A) 200 ppm (B) 300 ppm
(C) 400 ppm (D) 500 ppm

1577. Which of the following is a complex morphological unit?
(A) Fruit (B) Flower
(C) Vegetable (D) All of these

1578. Water disinfection is done by ....
(A) Sodium chloride (B) Sodium hypo chloride
(C) Sodium Bai chloride (D) Potassium chloride

1579. Bluing of Rose petals is due to ....
(A) Accumulation of Ammonia (B) Nitrogen deficiency
(C) Calcium deficiency (D) Boron deficiency

1580. Hypobaric storage is also known as ....
(A) Modified atmospheric storage (B) Controlled atmospheric storage
(C) Low pressure storage (D) Cold storage

1581. Limp neck is disorder of ....
(A) Rose (B) Tulip
(C) Dahlia (D) Gurhal

1582. Quilling of florets is a common disorder of..........
(A) Tuberose (B) Gladiolus
(C) Chrysanthemum (D) Dahlia

1583. Leaf scorching in Lily is due to .... deficiency
(A) Boron (B) Fluorine
(C) Manganese (D) Aluminium

1584. Which site is appropriate for Lawn in home garden?

(A) South (B) South East

(C) South West (D) All of these

1585. In one mowing of Lawn how much grass showed is cut?

(A) 1/2 (B) 1/3

(C) 1/4 (D) Any of the above

1586. Fairy ring spot in Lawn is due to ....

(A) Nutrient deficiency (B) Fungus

(C) Bacteria (D) Virus

1587. *Delonix regia* is commonly known as ....

(A) Blue Gulmohar (B) Red Gulmohar

(C) Yellow Gulmohar (D) All of these

1588. Kigellia has beautiful ....

(A) Flower (B) Foliage

(C) Fruits (D) Bracts

1589. Neem belong to family ....

(A) Bignoniaceae (B) Maliaceae

(C) Sapoteceae (D) Tilleaceae

1590. Which is/are fragrant flower tree?

(A) Magnolia (B) Devil's tree

(C) Michelia (D) All of these

1591. A shade tree should have .... canopy

(A) Wide oval (B) Wild

(D) Round (D) Pyramidal

1592. Which is/are white colour flower tree?

(A) Magnolia (B) Mimusops

(C) Plumeria (D) All of these

1593. Thuja is grown for ....

(A) Flower (B) Fruits

(C) Cut foliage (D) Edge

1594. Which species of *Ficus* is grown for religious purpose?

(A) *bengalensis* (B) *elastic*

(C) *religiosa* (D) *carica*

1595. Which is the common weed of lawn ....

(A) Motha (B) *Oxalis sp.*

(C) *Joysia japonica* (D) Parthenium

1596. Commonly used medium in air layering is ....

(A) Takakiopsida moss
(B) Ambuchanania moss
(C) Sphagnum moss
(D) Bryophyta moss

1597. Acalypha belongs to family ....

(A) Euphorbiaceae
(B) Leguminoceae
(C) Asteracea
(D) Verbenaceae

1598. Which of the following common rootstocks is popular in northern plains of India?

(A) *Rosa bourboniana*
(B) *Rosa laxa*
(C) *Rosa canina*
(D) *Rosa maneth*

1599. The ideal pH range of the soil for growing the lawn is ....

(A) 7.0-8.0
(B) 5.0-5.6
(C) 6.5-7.5
(D) 8.0-9.0

1600. Which of the following is active compound in Neem oil?

(A) Parathion
(B) Azadirachitin
(C) Rotenone
(D) Endrin

1601. The Persian, Italian and Mughal gardens are example of ....

(A) Informal garden
(B) Formal garden
(C) Wild garden
(D) Free style garden

1602. Main constituent of bordeaux mixture is ....

(A) Copper sulphate
(B) Zinc sulphate
(C) Bromide
(D) Iron

1603. Which of the following plants is suitable for ornamental internal hedges?

(A) *Caesalpinia pulcherrima*
(B) *Carissa carandas*
(C) *Cupressus macrocarpa*
(D) *Casuarina equisetifolia*

1604. NRC of Orchid is situated at ....

(A) Assam
(B) West Bengal
(C) Sikkim
(D) Nagaland

1605. Which of the ground cover plants belongs to sunflower family?

(A) Wedalia
(B) Acalypha
(C) Ipomea
(D) Verbena

1606. Which of the following is national flower of our country?

(A) Dahlia
(B) Rose
(C) Lotus
(D) Lilium

1607. The Government Botanical Garden, Udhagamandalam (Ooty) was established in ....

(A) 1878 (B) 1848

(C) 1982 (D) 1939

1608. The colour of *Portulaca grandifiora* flower is ....

(A) White (B) Various

(C) Pink (D) Red

1609. Stone lanterns are important feature in .... gardens.

(A) Japanese (B) Mughal

(C) Persian (D) English

1610. Botanical name of bermuda grass is ....

(A) *Buchloe dactyloides* (B) *Cynodon dactylon*

(C) *Axonopus affinis* (D) *Eremochloa ophiuroides*

1611. 'Swarna Rekha' is a gamma ray induced mutant cultivar of ....

(A) Gladiolus (B) Tuberose

(C) Dahlia (D) Rose

1612. Rose garden of Ludhiana is example of ....

(A) Formal garden (B) Informal garden

(C) Free style garden (D) Wild garden

1613. Which is used for rising in hanging baskets?

(A) Bigonia (B) Fabaceae

(C) Petunoideae (D) Caryophyllaceae

1614. Which is the permanent bulb suitable for growing well indoors year after year?

(A) Amaryllis (B) Tulip

(C) Hyacinths (D) Narcissus

1615. CO_2 requirement of chrysanthemums grown under polyhouses is .... ppm

(A) 100-200 (B) 600-900

(C) 500-800 (D) 1000-1200

1616. First flower show at Udhagamandalam was organized in ....

(A) 1895 (B) 1890

(C) 1943 (D) 1727

1617. An area where seedlings, cuttings and grafts are raised with care before transplanting?

(A) Garden (B) Nursery

(C) Park (D) Orchard

1618. Which of the following rose sp. is grown up for extracting more essential oil?
   (A) *R. gallica* (B) *R. canina*
   (C) *R. damascene* (D) *R. sericea*

1619. Botanically, plants which have the special structure to climb on supports are defined as....
   (A) Edge (B) Hedge
   (C) Climbers (D) Creepers

1620. The most common used hormone for rooting is ....
   (A) Ethylene (B) GA3
   (C) IBA (D) Cytokinin

1621. Lal Bagh Botanical Garden is located at ....
   (A) Kolkata (B) Bengaluru
   (C) Darjeeling (D) New Delhi

1622. Which methods of lawn making is the cheapest but time-consuming?
   (A) Dibbling roots (B) Seedling
   (C) Turfing (D) Turf plastering

1623. Which of the following is the most successful method for multiplying roses?
   (A) Grafting (B) Budding
   (C) Ground layering (D) Air layering

1624. Economic part of the Glorylily (Gloriosa) is ....
   (A) Seed (B) Flower
   (C) Stem (D) Leaf

1625. Garden of Fatehpur Sikri was laid out by the king ....
   (A) Fadai Khan (B) Nurjhan
   (C) Akbar (D) Jahangir

1626. Green Highways policy was launched in ....
   (A) 2015 (B) 2012
   (C) 2016 (D) 2018

1627. Which of the following plant is not suitable for a topiary?
   (A) *Celosia spp.* (B) *Caesalphinia coriari*
   (C) *Jaminum sp.* (D) *Bougainvillea sp.*

1628. Which of the following trees adorns with scented white bloom?
   (A) *Adambea glabra Lam* (B) *Mimusops elengi*
   (C) *Cassia roxburghii* (D) *Cassia javanica*

1629. Which of the following young plants are highly sensitive to chilling injury?
   (A) Anthurium (B) Freesia
   (C) Carnation (D) Tulip

1630. Delhi Crossandra is a .... Species

(A) Diploid (B) Triploid
(C) Tetraploid (D) Hexaploid

1631. National flower of Turkey is ....

(A) Dahlia (B) Tulip
(C) Bird of Paradise (D) Amaranths

1632. Bird of Paradise originated from ....

(A) South America (B) North America
(C) South Africa (D) Mexico

1633. Bird of Paradise is propagated through ....

(A) Seeds (B) Division of clumps
(C) Both a & b (B) Cuttings

1634. Storage temperature for Bird of Paradise is .... ºC

(A) 1-2 (B) 5
(C) 10 (D) 8

1635. Freesia is native to ....

(A) South America (B) North America
(C) South Africa (D) Mexico

1636. Basic chromosome number in Freesia is ....

(A) 7 (B) 10
(C) 9 (D) 11

1637. Freesia belong to family ....

(A) Iridaceae (B) Plumbaginaceae
(C) Fabaceae (D) Amaranthaceae

1638. Gloriosa belong to family ....

(A) Liliaceae (B) Plumbaginaceae
(C) Fabaceae (D) Amaranthaceae

1639. Planting season of Gloriosa is ....

(A) June-July (B) Sept-Oct
(C) Dec-Jan (D) March-April

1640. Basic chromosome number of Alstromeria is ....

(A) 7 (B) 8
(C) 9 (D) 11

1641. Alstromeria is native to ....

(A) Tropical America (B) India
(C) China (D) Mexico

1642. Blind shoot production increase in Alstromeria due to ....
(A) High temperature (B) Low temperature
(C) Insect attack (D) Short day

1643. Alstromeria is an .... plant
(A) Annual (B) Perennial
(C) Bie-Annual (D) Monthly

1644. Alstromeria flowers should be stored at .... °C with 80-90% RH
(A) 3-4 (B) 1-2
(C) 0-1 (D) 4-5

1645. Zantedeschia belong to family ....
(A) Araceae (B) Plumbaginaceae
(C) Fabaceae (D) Amaranthaceae

1646. Zantedeschia originated from ....
(A) South Africa (B) India
(C) China (D) Mexico

1647. Zantedeschia cut flowers can be stored at .... °C
(A) 0 (B) -2
(C) 1 (D) 5

1648. Rose variety Mrinalini has been released from ....
(A) IARI (B) IIHR
(C) DFR (D) PAU

1649. Priyadarshini variety of Rose belong to group ....
(A) Hybrid tea (B) Floribundas
(C) Grandiflora (D) Barboniana

1650. Madhosh variety of Rose released from ....
(A) IARI (B) IIHR
(C) DFR (D) PAU

1651. Dhiraj variety of Gladiolus has been developed from ....
(A) NBRI (B) IIHR
(C) MPKV (D) FRI

1652. Shubangini is a variety of ....
(A) Rose (B) Gladiolus
(C) Tuberose (D) Chrysanthemum

1653. Fidelo is a variety of ....
(A) Rose (B) Gladiolus
(C) Tuberose (D) Chrysanthemum

1654. Shakti is a .... Variety of Gladiolus

(A) Induced mutant (B) Hybrid
(C) Introduced (D) Chemical mutant

1655. Shobha variety of Gladiolus has been released from ....

(A) NBRI (B) IIHR
(C) MPKV (D) FRI

1656. .... is created in any garden to avoid monotonous view.

(A) Accent (B) Emphasis
(C) Proportion (D) Both A & B

1657. The overall effect of various features is known as ....

(A) Harmony (B) Trophy
(C) Scale (D) Unity

1658. The total mass of any object or plant in landscape design is termed as ....

(A) Balance (B) Proportion
(C) Unity (D) Line

1659. Maximum somatic hybrids have been developed in ....

(A) Petunia (B) Geranium
(C) Stock (D) None of above

1660. A three dimensional confined view of terminal building is known as

(A) Axis (B) Vista
(C) Circulation (D) None of above

1661. Which of the following is an element of landscape gardening

(A) Pattern (B) Form
(C) Colour (D) All of above.

1662. Which of the following is a neutral colour in colour scheme

(A) White (B) Grey
(C) Red (D) Both a & b

1663. Tertiary colour are formed by mixing of

(A) Red and white (B) Green and orange
(C) Red and Blue (D) None of above.

1664. Which of the following annuals are used as cut flowers?

(A) Carnation (B) Corn flower
(C) Sweet William (D) All of these

1665. Which of the following annuals are used for making dry flower products?

(A) Statice (B) Helichrysum
(C) Nigella (D) All of above

1666. According to photoperiodic requirement Amaranths is a ....

(A) Long day plant (B) Short day plant

(C) Day neutral (D) Long short day

1667. Bougainvillea climbs with the help of special structure called ....

(A) Tendrils (B) Thorns

(C) Roots (D) Sticky substances

1668. An example of rambler is ....

(A) Lonicera (B) *Ficus repens*

(C) *Quisqualis indica* (D) Bigonia

1669. Which of the following climber has fragrant flowers?

(A) Jasmine (B) Honey suckle

(C) Both a & b (D) None of above

1670. Which of the following climber can grow under shade?

(A) *Clerodendron splendens* (B) *Thunbergia grandiflora*

(C) *Jacquemontia pentantha* (D) All of above

1671. Which of the following is a climber?

(A) *Ficus elastic* (B) *Ficus infectoria*

(C) *Ficus repens* (D) *Ficus religiosa*

1672. Gomphrena is a

(A) Long day plant (B) Short long day plant

(C) Intermediate plant (D) Day neutral plant

1673. Which of the following annual is used for screening purpose?

(A) Hollyhock (B) Sweet Pea

(C) Moon flower (D) All of above.

1674. Marigold is propagated through ....

(A) Cutting (B) Seed

(C) Tissue culture (D) All of above

1675. Rose is commercially propagated through ....

(A) Cutting (B) T budding

(C) Seed (D) Suckers

1676. Patch budding is a propagation method of .... tree.

(A) *Abelia xgrandiflora* (B) Pecan

(C) Siberian peashrub (D) All of above

1677. Dahlia is propagating by .... cuttings.

(A) Stem cuttings (B) Tuberous cutting

(C) Root cutting (D) Leaf cutting

1678. .... is propagating by mound or stool layering.

(A) Rose (B) Jasmine
(C) Marigold (D) Gerbera

1679. Which of the following flower crops are propagated through seed

(A) Chrysanthemum (B) Rose
(C) Tuberose (D) None of above

1680. Rhododendron is propagated through

(A) Saddle grafting (B) Inarch grafting
(C) T-budding (D) All of the above

1681. Flat grafting is most commonly used for propagation of ....

(A) Bulbous plants (B) Trees
(C) Cacti and succulents (D) Both a & c

1682. Which of the following is a summer annual?

(A) Salvia (B) Sweet Pea
(C) Kochia (D) Sweet alyssum

1683. *Magnolia grandiflora* is propagated through

(A) Seeds (B) Cutting
(C) Air layering (D) All of above.

1684. F1 hybrid seed of petunia was first produced by .... Pvt.Ltd.

(A) Indian hybrid seed company (B) Indo-American hybrid seed company
(C) Maharashtra seed company (D) None of these

1685. Which is an examples of self-pollinated seasonal flower?

(A) China Aster (B) Alstroemeria
(C) Tuberose (D) Heliconia

1686. The pH of soil for seed production should be ....

(A) 5.5-7.0 (B) 5-7
(C) 6-7 (D) 4-5

1687. Which of the following is used to checked pollen sterility in rose?

(A) Ethyl alcohol (B) Potassium iodide
(C) Acetocarmine (D) All of these

1688. Which of the following has genetic male sterility?

(A) Marigold (B) Zinnia
(C) Ageratum (D) All of above

1689. Which of the following has self-incompatibility

(A) Petunia (B) Pansy
(C) Both a and b (D) None of above

1690. In which colour Pelargenidin anthocyanidin is present?
(A) Orange-red (B) Purple
(C) Green (D) Blue

1691. Which gene action can control the inheritance of pigments?
(A) Epastasis (B) Additive
(C) Dominance (D) All of above

1692. Which of the following is an often cross pollinated flower crop
(A) Antirrhinum (B) Aster
(C) Dahlia (D) All of above.

1693. Which elements are used for building up the plant architecture?
(A) CHO (B) NPK
(C) Nitrogen (D) Ca, Mg, S

1694. The criteria of essentiality for nutrient was proposed by ....
(A) Lysenko (B) Robert Browen
(C) Arnon and Stout (D) M.S. Swaminathan

1695. How much % nitrogen is found in whole plant.
(A) 2 % (B) 3 %
(C) 4 % (D) 6 %

1696. The interveinal chlorosis in carnation is a symptom of ....
(A) 'Fe' deficiency (B) 'B' deficiency
(C) 'Mn' deficiency (D) All of above

1697. Sebaculis red is variety of ....
(A) Tuberose (B) Marigold
(C) Jasmine (D) Crossandra

1698. Pale foliage in rose is a deficiency symptom of ....
(A) Fe (B) Mn
(C) B (D) Cu

1699. Phosphorous is available in which of the following form?
(A) $H_2PO_4^-$ (B) HPO_4^{2-}
(C) Both a & b (D) None of above

1700. The total quantity of boron in plants is found as
(A) 25-75 ppm (B) 10-1500 ppm
(C) 3-150 ppm (D) 5-1500 ppm.

1701. In Ikebana Shine is placed at the angel of ....
(A) 45 (B) 75
(C) 15 (D) 30

1702. Ikebana Shoe is placed at the angel of ....

(A) 45 (B) 75
(C) 15 (D) 30

1703. Ikebana Hikae is placed at the angel of ....

(A) 45 (B) 75
(C) 15 (D) 30

1704. Rose mutant Abhisarka is developed from ....

(A) Kiss of fire (B) Gulzar
(C) Banjaran (D) Arjun

1705. For cascade type bonsai which type container is suitable?

(A) Square (B) Oval
(C) Round (D) Rectangular

1706. What is the ideal PH for rose growing?

(A) 6-7.5 (B) 3-4.5
(C) 4-5 (D) 7.5-8.5

1707. Wild garden developed by William Robinson in .... century

(A) 18th (B) 19TH
(C) 20TH (D) 17TH

1708. Devil's tree is also known as..........

(A) Saptaprani (B) Amaltas
(C) Arjun (D) None of above

1709. *Alstonia scholaris* is botanical name of ....

(A) Arjun (B) Balam khira
(C) Devil's tree (D) Champa

1710. Mame bonsai height should not be more than .... cm

(A) 2-5 (B) 5-15
(C) 15-20 (D) 20-30

1711. How many flowers produced in double type Gerbera in ployhouse in per square metre?

(A) 400 (B) 250
(C) 300 (D) 200

1712. *Aracuria cookii* is highly suitable for ....

(A) Hedge (B) Edge
(C) Specimen (D) Screening

1713. Ascocenda is the hybrid of which flower?

(A) Gerbera (B) Orchids
(C) Gladiolus (D) Tuberose

1714. Which of the following is most efficient irrigation method for flower crops?
(A) Sprinkler (B) Drip
(C) Flood (D) Trench

1715. Dimorphotheca is very good for ....
(A) Hanging baskets (B) Bedding
(C) Pots (D) Bothc b & c

1716. Rose variety Mrinalini released from ....
(A) IARI (B) IIHR
(C) NBRI (D) PAU

1717. Which fungicide is used to prevent rose plant from dieback?
(A) Rose paint (B) Bavistin
(C) Captan (D) Zineb

1718. Which type Chrysanthemum is used for garlands and veni bracelets?
(A) Standard (B) Spray
(C) Pompon (D) Posy

1719. Central institute of medicinal and aromatic plants is located at ....
(A) Lucknow (B) New Delhi
(C) Gurgaon (D) Punjab

1720. *Alyssum maritimum* is an .... blooming annual
(A) Early (B) Late
(C) Mid (D) Very late

1721. Which type of gardening system was developed by Dr. Nathanial Ward?
(A) Hanging baskets (B) Vertical
(C) Terrariums (D) Jar and bottle gardening

1722. French Marigold is an ....
(A) Diploid (B) Tetraploid
(C) Triploid (D) Hexaploid

1723. Which colour flower bears in *Antigonon leptopus*?
(A) Red (B) Pink
(C) White (D) Blue

1724. Kamini and Shashank are the verities of ....
(A) Tuberose (B) China aster
(C) Marigold (D) Rose

1725. Leading flowers seed producing state in India?
(A) Haryana (B) Tamil Nadu
(C) West Bengal (D) Punjab

1726. Portulaca cultivar Jhumka is developed at?

(A) IIHR (B) NBRI

(C) PAU (D) IARI

1727. Yellow is contrasting to which colour?

(A) Orange (B) Violet

(C) Blue (D) Red

1728. Pseudo bulbs are found in which flower crop?

(A) Tuberose (B) Canna

(C) Orchid (D) All of these

1729. Gladiolus is propagated by?

(A) Bulb (B) Tuber

(C) Corm (D) Rhizome

1730. Swollen end of stolon is known as?

(A) Tuber (B) Bulbils

(C) Cormel (D) None of these

ANSWER KEY

1. (A)	2. (C)	3. (B)	4. (C)	5. (B)	6. (D)	7. (B)	8. (C)
9. (C)	10. (A)	11. (C)	12. (B)	13. (C)	14. (A)	15. (B)	16. (B)
17. (D)	18. (C)	19. (B)	20. (A)	21. (C)	22. (C)	23. (C)	24. (B)
25. (A)	26. (A)	27. (A)	28. (A)	29. (A)	30. (A)	31. (A)	32. (A)
33. (A)	34. (A)	35. (A)	36. (A)	37. (A)	38. (A)	39. (A)	40. (A)
41. (A)	42. (A)	43. (A)	44. (A)	45. (A)	46. (A)	47. (A)	48. (A)
49. (A)	50. (A)	51. (A)	52. (A)	53. (A)	54. (A)	55. (A)	56. (A)
57. (B)	58. (C)	59. (C)	60. (D)	61. (A)	62. (B)	63. (A)	64. (B)
65. (B)	66. (B)	67. (C)	68. (D)	69. (A)	70. (B)	71. (A)	72. (C)
73. (B)	74. (B)	75. (B)	76. (C)	77. (D)	78. (C)	79. (D)	80. (B)
81. (C)	82. (C)	83. (C)	84. (B)	85. (B)	86. (C)	87. (C)	88. (A)
89. (B)	90. (A)	91. (B)	92. (B)	93. (C)	94. (B)	95. (B)	96. (C)
97. (B)	98. (D)	99. (B)	100. (B)	101. (B)	102. (C)	103. (D)	104. (A)
105. (C)	106. (C)	107. (B)	108. (B)	109. (B)	110. (A)	111. (C)	112. (A)
113. (A)	114. (B)	115. (B)	116. (D)	117. (B)	118. (B)	119. (A)	120. (A)
121. (C)	122. (B)	123. (B)	124. (B)	125. (B)	126. (B)	127. (A)	128. (A)
129. (C)	130. (D)	131. (D)	132. (C)	133. (A)	134. (A)	135. (C)	136. (A)
137. (A)	138. (B)	139. (A)	140. (D)	141. (B)	142. (A)	143. (A)	144. (A)
145. (B)	146. (B)	147. (A)	148. (C)	149. (B)	150. (A)	151. (B)	152. (A)
153. (A)	154. (D)	155. (B)	156. (A)	157. (C)	158. (C)	159. (A)	160. (A)
161. (C)	162. (B)	163. (A)	164. (D)	165. (A)	166. (C)	167. (A)	168. (B)
169. (B)	170. (D)	171. (C)	172. (A)	173. (B)	174. (A)	175. (D)	176. (A)
177. (D)	178. (A)	179. (D)	180. (D)	181. (A)	182. (A)	183. (A)	184. (B)
185. (C)	186. (B)	187. (C)	188. (D)	189. (C)	190. (D)	191. (B)	192. (C)
193. (B)	194. (C)	195. (C)	196. (C)	197. (B)	198. (B)	199. (D)	200. (D)
201. (D)	202. (B)	203. (C)	204. (B)	205. (B)	206. (A)	207. (A)	208. (B)
209. (B)	210. (D)	211. (C)	212. (B)	213. (B)	214. (A)	215. (B)	216. (C)
217. (D)	218. (C)	219. (A)	220. (B)	221. (B)	222. (B)	223. (C)	224. (B)
225. (B)	226. (D)	227. (C)	228. (B)	229. (B)	230. (C)	231. (A)	232. (C)
233. (A)	234. (B)	235. (B)	236. (B)	237. (B)	238. (B)	239. (B)	240. (A)
241. (D)	242. (B)	243. (C)	244. (C)	245. (C)	246. (B)	247. (B)	248. (B)
249. (C)	250. (A)	251. (B)	252. (B)	253. (B)	254. (B)	255. (A)	256. (C)
257. (B)	258. (C)	259. (B)	260. (D)	261. (B)	262. (B)	263. (B)	264. (C)
265. (B)	266. (D)	267. (A)	268. (D)	269. (C)	270. (C)	271. (C)	272. (D)
273. (A)	274. (A)	275. (B)	276. (B)	277. (B)	278. (D)	279. (C)	280. (D)
281. (B)	282. (A)	283. (C)	284. (A)	285. (C)	286. (B)	287. (C)	288. (B)

289. (B)	290. (B)	291. (C)	292. (C)	293. (C)	294. (A)	295. (B)	296. (B)
297. (A)	298. (B)	299. (A)	300. (C)	301. (B)	302. (A)	303. (C)	304. (B)
305. (A)	306. (B)	307. (C)	308. (B)	309. (D)	310. (C)	311. (D)	312. (A)
313. (A)	314. (A)	315. (D)	316. (A)	317. (B)	318. (A)	319. (A)	320. (C)
321. (D)	322. (A)	323. (D)	324. (A)	325. (C)	326. (D)	327. (A)	328. (A)
329. (A)	330. (C)	331. (B)	332. (B)	333. (B)	334. (C)	335. (A)	336. (C)
337. (C)	338. (B)	339. (C)	340. (B)	341. (D)	342. (B)	343. (D)	344. (D)
345. (B)	346. (D)	347. (B)	348. (D)	349. (A)	350. (A)	351. (D)	352. (D)
353. (A)	354. (A)	355. (D)	356. (D)	357. (A)	358. (B)	359. (D)	360. (B)
361. (B)	362. (B)	363. (A)	364. (C)	365. (B)	366. (A)	367. (D)	368. (C)
369. (B)	370. (B)	371. (B)	372. (B)	373. (B)	374. (B)	375. (B)	376. (D)
377. (C)	378. (B)	379. (B)	380. (B)	381. (C)	382. (C)	383. (D)	384. (C)
385. (D)	386. (C)	387. (B)	388. (D)	389. (A)	390. (B)	391. (C)	392. (C)
393. (B)	394. (C)	395. (C)	396. (B)	397. (D)	398. (C)	399. (B)	400. (B)
401. (A)	402. (C)	403. (C)	404. (A)	405. (C)	406. (B)	407. (B)	408. (D)
409. (C)	410. (B)	411. (C)	412. (A)	413. (B)	414. (B)	415. (C)	416. (B)
417. (B)	418. (B)	419. (B)	420. (D)	421. (C)	422. (B)	423. (D)	424. (B)
425. (C)	426. (C)	427. (C)	428. (A)	429. (C)	430. (D)	431. (B)	432. (B)
433. (A)	434. (C)	435. (B)	436. (B)	437. (B)	438. (C)	439. (B)	440. (C)
441. (B)	442. (C)	443. (B)	444. (C)	445. (C)	446. (D)	447. (C)	448. (C)
449. (C)	450. (B)	451. (C)	452. (C)	453. (B)	454. (B)	455. (B)	456. (B)
457. (D)	458. (A)	459. (B)	460. (B)	461. (D)	462. (D)	463. (D)	464. (B)
465. (A)	466. (D)	467. (A)	468. (C)	469. (C)	470. (A)	471. (B)	472. (B)
473. (C)	474. (B)	475. (D)	476. (B)	477. (B)	478. (C)	479. (D)	480. (A)
481. (B)	482. (C)	483. (B)	484. (B)	485. (C)	486. (B)	487. (B)	488. (C)
489. (B)	490. (C)	491. (D)	492. (D)	493. (C)	494. (A)	495. (A)	496. (D)
497. (A)	498. (C)	499. (D)	500. (C)	501. (B)	502. (A)	503. (B)	504. (B)
505. (B)	506. (A)	507. (A)	508. (B)	509. (C)	510. (D)	511. (B)	512. (A)
513. (B)	514. (B)	515. (C)	516. (D)	517. (C)	518. (C)	519. (B)	520. (A)
521. (A)	522. (A)	523. (C)	524. (A)	525. (D)	526. (A)	527. (B)	528. (B)
529. (C)	530. (C)	531. (B)	532. (A)	533. (C)	534. (B)	535. (A)	536. (C)
537. (B)	538. (C)	539. (D)	540. (C)	541. (B)	542. (C)	543. (A)	544. (C)
545. (A)	546. (D)	547. (C)	548. (C)	549. (B)	550. (B)	551. (B)	552. (D)
553. (B)	554. (B)	555. (B)	556. (D)	557. (B)	558. (B)	559. (A)	560. (A)
561. (C)	562. (D)	563. (B)	564. (A)	565. (B)	566. (C)	567. (C)	568. (A)
569. (C)	570. (D)	571. (B)	572. (A)	573. (D)	574. (A)	575. (C)	576. (A)
577. (A)	578. (B)	579. (B)	580. (B)	581. (B)	582. (A)	583. (B)	584. (A)
585. (B)	586. (A)	587. (B)	588. (B)	589. (C)	590. (D)	591. (D)	592. (B)

593. (D)	594. (B)	595. (C)	596. (B)	597. (C)	598. (C)	599. (A)	600. (B)
601. (A)	602. (C)	603. (A)	604. (C)	605. (A)	606. (C)	607. (A)	608. (B)
609. (B)	610. (C)	611. (A)	612. (B)	613. (C)	614. (B)	615. (B)	616. (C)
617. (A)	618. (B)	619. (C)	620. (A)	621. (D)	622. (B)	623. (C)	624. (C)
625. (A)	626. (B)	627. (A)	628. (B)	629. (A)	630. (C)	631. (A)	632. (B)
633. (D)	634. (C)	635. (D)	636. (D)	637. (A)	638. (B)	639. (A)	640. (D)
641. (B)	642. (B)	643. (C)	644. (B)	645. (C)	646. (A)	647. (A)	648. (C)
649. (A)	650. (C)	651. (D)	652. (B)	653. (C)	653. (A)	655. (C)	656. (A)
657. (D)	658. (C)	659. (B)	660. (A)	661. (C)	662. (D)	663. (A)	664. (A)
665. (B)	666. (A)	667. (A)	668. (A)	669. (C)	670. (A)	671. (B)	672. (D)
673. (D)	674. (A)	675. (A)	676. (B)	677. (B)	678. (A)	679. (B)	680. (A)
681. (A)	682. (A)	683. (C)	684. (B)	685. (A)	686. (B)	687. (B)	688. (C)
689. (B)	690. (B)	691. (C)	692. (A)	693. (C)	693. (C)	695. (A)	696. (A)
697. (D)	698. (D)	699. (B)	700. (B)	701. (A)	702. (C)	703. (B)	704. (B)
705. (A)	706. (A)	707. (D)	708. (B)	709. (B)	710. (B)	711. (C)	712. (C)
713. (B)	714. (B)	715. (C)	716. (C)	717. (D)	718. (C)	719. (B)	720. (C)
721. (A)	722. (C)	723. (A)	724. (A)	725. (A)	726. (D)	727. (B)	728. (B)
729. (D)	730. (D)	731. (D)	732. (D)	733. (B)	734. (B)	735. (C)	736. (A)
737. (A)	738. (D)	739. (A)	740. (A)	741. (B)	742. (C)	743. (A)	744. (C)
745. (A)	746. (D)	747. (A)	748. (A)	749. (B)	750. (C)	751. (B)	752. (A)
753. (C)	754. (D)	755. (C)	756. (D)	757. (A)	758. (B)	759. (A)	760. (B)
761. (D)	762. (D)	763. (D)	764. (D)	765. (D)	766. (B)	767. (B)	768. (D)
769. (A)	770. (A)	771. (C)	772. (A)	773. (D)	773. (B)	775. (B)	776. (B)
777. (A)	778. (B)	779. (D)	780. (A)	781. (C)	782. (C)	783. (D)	784. (B)
785. (A)	786. (A)	787. (C)	788. (D)	789. (A)	790. (A)	791. (B)	792. (D)
793. (A)	794. (C)	795. (C)	796. (C)	797. (D)	798. (A)	799. (C)	800. (A)
801. (B)	802. (A)	803. (B)	804. (A)	805. (A)	806. (A)	807. (B)	808. (C)
809. (D)	810. (B)	811. (A)	812. (B)	813. (C)	814. (C)	815. (C)	816. (A)
817. (D)	818. (D)	819. (A)	820. (D)	821. (C)	822. (D)	823. (C)	824. (A)
825. (A)	826. (A)	827. (A)	828. (B)	829. (A)	830. (C)	831. (B)	832. (B)
833. (B)	834. (B)	835. (C)	836. (C)	837. (B)	838. (A)	839. (D)	840. (C)
841. (C)	842. (C)	843. (B)	844. (C)	845. (B)	846. (A)	847. (A)	848. (D)
849. (D)	850. (D)	851. (B)	852. (B)	853. (D)	853. (D)	855. (B)	856. (C)
857. (D)	858. (A)	859. (C)	860. (B)	861. (A)	862. (D)	863. (D)	864. (A)
865. (B)	866. (C)	867. (C)	868. (B)	869. (B)	870. (C)	871. (C)	872. (B)
873. (C)	874. (C)	875. (C)	876. (A)	877. (D)	878. (D)	879. (D)	880. (B)
881. (B)	882. (B)	883. (D)	884. (D)	885. (B)	886. (C)	887. (B)	888. (D)
889. (D)	890. (B)	891. (A)	892. (B)	893. (D)	894. (D)	895. (A)	896. (C)

897. (A)	898. (C)	899. (B)	900. (B)	901. (B)	902. (A)	903. (D)	904. (A)
905. (C)	906. (A)	907. (B)	908. (A)	909. (A)	910. (B)	911. (B)	912. (B)
913. (C)	914. (C)	915. (A)	916. (A)	917. (A)	918. (A)	919. (B)	920. (B)
921. (B)	922. (A)	923. (B)	923. (D)	925. (A)	926. (B)	927. (C)	928. (B)
929. (A)	930. (C)	931. (A)	932. (B)	933. (B)	934. (B)	935. (B)	936. (D)
937. (B)	938. (B)	939. (C)	940. (B)	941. (B)	942. (A)	943. (A)	944. (B)
945. (D)	946. (C)	947. (D)	948. (C)	949. (A)	950. (B)	951. (C)	952. (B)
953. (B)	954. (B)	955. (B)	956. (BQ)	957. (B)	958. (B)	959. (A)	960. (B)
961. (B)	962. (D)	963. (C)	964. (B)	965. (A)	966. (B)	967. (B)	968. (C)
969. (A)	970. (A)	971. (B)	972. (A)	973. (C)	974. (B)	975. (B)	976. (D)
977. (B)	978. (B)	979. (A)	980. (B)	981. (B)	982. (A)	983. (A)	984. (A)
985. (B)	986. (A)	987. (A)	988. (B)	989. (B)	990. (B)	991. (B)	992. (B)
913. (A)	994. (B)	995. (B)	996. (B)	997. (B)	998. (C)	999. (B)	1000. (B)
1001. (C)	1002. (C)	1003. (A)	1003. (C)	1005. (B)	1006. (D)	1007. (C)	1008. (C)
1009. (C)	1010. (A)	1011. (A)	1012. (B)	1013. (C)	1014. (C)	1015. (C)	1016. (C)
1017. (C)	1018. (A)	1019. (D)	1020. (B)	1021. (B)	1022. (B)	1023. (C)	1024. (D)
1025. (A)	1026. (C)	1027. (D)	1028. (C)	1029. (C)	1030. (C)	1031. (C)	1032. (B)
1033. (B)	1034. (C)	1035. (C)	1036. (A)	1037. (B)	1038. (A)	1039. (A)	1040. (B)
1041. (D)	1042. (B)	1043. (B)	1044. (A)	1045. (C)	1046. (A)	1047. (A)	1048. (B)
1049. (A)	1050. (C)	1051. (B)	1052. (B)	1053. (B)	1054. (B)	1055. (B)	1056. (A)
1057. (C)	1058. (A)	1059. (A)	1060. (A)	1061. (A)	1062. (A)	1063. (B)	1064. (B)
1065. (C)	1066. (A)	1067. (A)	1068. (A)	1069. (B)	1070. (A)	1071. (A)	1072. (C)
1073. (B)	1074. (A)	1075. (B)	1076. (A)	1077. (C)	1078. (B)	1079. (D)	1080. (C)
1081. (A)	1082. (A)	1083. (B)	1083. (A)	1085. (A)	1086. (A)	1087. (B)	1088. (A)
1089. (B)	1090. (D)	1091. (B)	1092. (C)	1093. (C)	1094. (B)	1095. (B)	1096. (B)
1097. (C)	1098. (D)	1099. (C)	1100. (A)	1101. (B)	1102. (B)	1103. (B)	1104. (B)
1105. (B)	1106. (C)	1107. (B)	1108. (B)	1109. (B)	1110. (B)	1111. (C)	1112. (B)
1113. (B)	1114. (B)	1115. (B)	1116. (C)	1117. (C)	1118. (A)	1119. (B)	1120. (B)
1121. (C)	1122. (B)	1123. (C)	1124. (B)	1125. (A)	1126. (B)	1127. (A)	1128. (A)
1129. (A)	1130. (A)	1131. (B)	1132. (B)	1133. (B)	1134. (A)	1135. (B)	1136. (B)
1137. (B)	1138. (B)	1139. (B)	1140. (A)	1141. (D)	1142. (B)	1143. (B)	1144. (B)
1145. (C)	1146. (B)	1147. (B)	1148. (A)	1149. (A)	1150. (C)	1151. (A)	1152. (B)
1153. (B)	1154. (C)	1155. (B)	1156. (A)	1157. (C)	1158. (A)	1159. (A)	1160. (B)
1161. (B)	1162. (A)	1163. (A)	1164. (B)	1165. (B)	1166. (A)	1167. (B)	1168. (A)
1169. (A)	1170. (B)	1171. (B)	1172. (C)	1173. (A)	1174. (C)	1175. (A)	1176. (A)
1177. (B)	1178. (C)	1179. (A)	1180. (A)	1181. (B)	1182. (A)	1183. (A)	1184. (A)
1185. (A)	1186. (B)	1187. (A)	1188. (C)	1189. (A)	1190. (B)	1191. (A)	1192. (D)
1193. (A)	1194. (B)	1195. (A)	1196. (C)	1197. (B)	1198. (D)	1199. (C)	1200. (B)

1201. (B)	1202. (D)	1203. (A)	1204. (A)	1205. (B)	1206. (A)	1207. (A)	1208. (C)
1209. (B)	1210. (B)	1211. (B)	1212. (D)	1213. (B)	1214. (B)	1215. (C)	1216. (C)
1217. (C)	1218. (C)	1219. (A)	1220. (A)	1221. (B)	1222. (C)	1223. (C)	1224. (C)
1225. (C)	1226. (C)	1227. (C)	1228. (B)	1229. (C)	1230. (C)	1231. (C)	1232. (B)
1233. (C)	1234. (C)	1235. (A)	1236. (D)	1237. (B)	1238. (A)	1239. (B)	1240. (D)
1241. (B)	1242. (C)	1243. (C)	1244. (D)	1245. (A)	1246. (B)	1247. (A)	1248. (B)
1249. (B)	1250. (B)	1251. (C)	1252. (D)	1253. (A)	1254. (B)	1255. (A)	1256. (C)
1257. (B)	1258. (B)	1259. (B)	1260. (C)	1261. (D)	1262. (C)	1263. (D)	1264. (B)
1265. (C)	1266. (C)	1267. (C)	1268. (B)	1269. (B)	1270. (C)	1271. (C)	1272. (A)
1273. (B)	1274. (A)	1275. (B)	1276. (A)	1277. (C)	1278. (B)	1279. (B)	1280. (D)
1281. (B)	1282. (D)	1283. (B)	1284. (B)	1285. (B)	1286. (A)	1287. (A)	1288. (B)
1289. (B)	1290. (B)	1291. (B)	1292. (A)	1293. (C)	1294. (A)	1295. (A)	1296. (C)
1297. (A)	1298. (B)	1299. (B)	1300. (C)	1301. (C)	1302. (C)	1303. (B)	1304. (C)
1305. (C)	1306. (A)	1307. (A)	1308. (B)	1309. (A)	1310. (A)	1311. (C)	1312. (A)
1313. (B)	1314. (D)	1315. (A)	1316. (C)	1317. (C)	1318. (C)	1319. (A)	1320. (B)
1321. (B)	1322. (A)	1323. (C)	1324. (A)	1325. (D)	1326. (A)	1327. (B)	1328. (C)
1329. (A)	1330. (B)	1331. (A)	1332. (B)	1333. (C)	1334. (B)	1335. (D)	1336. (A)
1337. (D)	1338. (A)	1339. (C)	1340. (D)	1341. (A)	1342. (A)	1343. (A)	1344. (C)
1345. (B)	1346. (B)	1347. (C)	1348. (A)	1349. (C)	1350. (C)	1351. (B)	1352. (C)
1353. (B)	1354. (D)	1355. (B)	1356. (A)	1357. (C)	1358. (B)	1359. (A)	1360. (B)
1361. (C)	1362. (B)	1363. (D)	1364. (C)	1365. (C)	1366. (C)	1367. (C)	1368. (A)
1369. (B)	1370. (B)	1371. (A)	1372. (B)	1373. (A)	1374. (C)	1375. (A)	1376. (C)
1377. (B)	1378. (B)	1379. (B)	1380. (B)	1381. (B)	1382. (B)	1383. (C)	1384. (B)
1385. (A)	1386. (C)	1387. (B)	1388. (A)	1389. (B)	1390. (C)	1391. (A)	1392. (A)
1393. (C)	1394. (D)	1395. (A)	1396. (C)	1397. (C)	1398. (A)	1399. (B)	1400. (A)
1401. (A)	1402. (B)	1403. (C)	1404. (A)	1405. (C)	1406. (B)	1407. (A)	1408. (A)
1409. (B)	1410. (A)	1411. (A)	1412. (C)	1413. (A)	1414. (B)	1415. (A)	1416. (A)
1417. (B)	1418. (B)	1419. (A)	1420. (B)	1421. (C)	1422. (A)	1423. (A)	1424. (A)
1425. (C)	1426. (A)	1427. (B)	1428. (B)	1429. (A)	1430. (A)	1431. (A)	1432. (B)
1433. (A)	1434. (B)	1435. (C)	1436. (B)	1437. (B)	1438. (C)	1439. (B)	1440. (A)
1441. (B)	1442. (B)	1443. (A)	1444. (B)	1445. (B)	1446. (C)	1447. (D)	1448. (A)
1449. (C)	1450. (C)	1451. (B)	1452. (B)	1453. (B)	1454. (A)	1455. (C)	1456. (A)
1457. (A)	1458. (B)	1459. (B)	1460. (D)	1461. (B)	1462. (B)	1463. (A)	1464. (A)
1465. (C)	1466. (B)	1467. (B)	1468. (B)	1469. (B)	1470. (B)	1471. (A)	1472. (A)
1473. (C)	1474. (D)	1475. (D)	1476. (C)	1477. (A)	1478. (A)	1479. (C)	1480. (A)
1481. (A)	1482. (B)	1483. (A)	1484. (D)	1485. (C)	1486. (A)	1487. (C)	1488. (B)
1489. (C)	1490. (A)	1491. (B)	1492. (B)	1493. (C)	1494. (A)	1495. (B)	1496. (B)
1497. (B)	1498. (C)	1499. (A)	1500. (A)	1501. (D)	1502. (A)	1503. (A)	1504. (A)

1505. (A)	1506. (B)	1507. (A)	1508. (A)	1509. (A)	1510. (C)	1511. (A)	1512. (A)
1513. (B)	1514. (A)	1515. (B)	1516. (C)	1517. (C)	1518. (C)	1519. (C)	1520. (C)
1521. (B)	1522. (A)	1523. (B)	1524. (B)	1525. (B)	1526. (A)	1527. (A)	1528. (B)
1529. (A)	1530. (A)	1531. (B)	1532. (C)	1533. (C)	1534. (B)	1535. (C)	1536. (B)
1537. (B)	1538. (A)	1539. (B)	1540. (C)	1541. (D)	1542. (C)	1543. (A)	1544. (B)
1545. (B)	1546. (B)	1547. (C)	1548. (B)	1549. (B)	1550. (D)	1551. (C)	1552. (B)
1553. (B)	1554. (C)	1555. (A)	1556. (C)	1557. (A)	1558. (B)	1559. (B)	1560. (B)
1561. (B)	1562. (B)	1563. (B)	1563. (A)	1565. (D)	1566. (B)	1567. (C)	1568. (C)
1569. (C)	1570. (B)	1571. (B)	1572. (B)	1573. (C)	1574. (A)	1575. (B)	1576. (B)
1577. (B)	1578. (B)	1579. (A)	1580. (C)	1581. (B)	1582. (C)	1583. (B)	1584. (D)
1585. (B)	1586. (B)	1587. (B)	1588. (D)	1589. (B)	1590. (D)	1591. (A)	1592. (D)
1593. (C)	1594. (C)	1595. (A)	1596. (C)	1597. (A)	1598. (A)	1599. (B)	1600. (B)
1601. (B)	1602. (A)	1603. (C)	1604. (C)	1605. (A)	1606. (C)	1607. (B)	1608. (B)
1609. (A)	1610. (B)	1611. (B)	1612. (C)	1613. (A)	1614. (A)	1615. (B)	1616. (A)
1617. (B)	1618. (C)	1619. (C)	1620. (C)	1621. (B)	1622. (A)	1623. (B)	1624. (A)
1625. (C)	1626. (A)	1627. (A)	1628. (B)	1629. (A)	1630. (B)	1631. (B)	1632. (C)
1633. (C)	1634. (C)	1635. (C)	1636. (B)	1637. (A)	1638. (A)	1639. (A)	1640. (B)
1641. (A)	1642. (A)	1643. (B)	1643. (B)	1645. (A)	1646. (A)	1647. (C)	1648. (A)
1649. (A)	1650. (A)	1651. (B)	1652. (B)	1653. (B)	1654. (A)	1655. (B)	1656. (D)
1657. (A)	1658. (B)	1659. (A)	1660. (B)	1661. (A)	1662. (D)	1663. (D)	1664. (D)
1665. (D)	1666. (B)	1667. (B)	1668. (D)	1669. (A)	1670. (A)	1671. (C)	1672. (D)
1673. (D)	1674. (B)	1675. (B)	1676. (B)	1677. (A)	1678. (B)	1679. (D)	1680. (A)
1681. (C)	1682. (C)	1683. (D)	1684. (B)	1685. (A)	1686. (A)	1687. (C)	1688. (D)
1689. (A)	1690. (A)	1691. (B)	1692. (D)	1693. (A)	1694. (C)	1695. (C)	1696. (C)
1697. (D)	1698. (A)	1699. (C)	1700. (A)	1701. (C)	1702. (A)	1703. (B)	1704. (A)
1705. (D)	1706. (A)	1707. (B)	1708. (A)	1709. (C)	1710. (B)	1711. (D)	1712. (C)
1713. (B)	1714. (B)	1715. (D)	1716. (A)	1717. (A)	1718. (B)	1719. (A)	1720. (A)
1721. (C)	1722. (B)	1723. (B)	1724. (B)	1725. (D)	1726. (B)	1727. (B)	1728. (C)
1729. (C)	1730. (C)						

MATCH THE FOLLOWINGS

(1) Shan Jahan	(A) Pinjor garden
(2) Jahangir	(B) Red Fort
(3) Akbar	(C) Salimar garden
(4) Fadai khan	(D) Fatapur garden

(1) B, (2) C, (3) D, (4) A

(1) Royal botanical garden	(A) Lucknow
(2) Govt. botanical garden	(B) Culcutta
(3) Royal Agri.- Horticulture society garden	(C) Ooty
(4) National botanical garden	(D) Kodaikanal
(5) Bryant's park	(E) Darjeeling

(1) E, (2) C, (3) B, (4) A, (5) D

(1) Potato creeper	(A) Asteraceae
(2) Money Plant	(B) Solanaceae
(3) Vernonia	(C) Verbanaceae
(4) Purple wreath	(D) Convolvulaceae
(5) Bridal Climber	(E) Araceae

(1) B, (2) E, (3) A, (4) C, (5) D

(1) Cat's claw	(A) Combretaceae
(2) Glory lilly	(B) Convolvulaceae
(3) Railway Creeper	(C) Araceae
(4) Australian Pineapple	(D) Liliaceae
(5) Rangoon Creeper	(E) Bignoniaceae

(1) E, (2) D, (3) B, (4) C, (5) A

(1) Coral Vine	(A) Aristolochiaceae
(2) Flame Vine	(B) Verbanaceae
(3) Duck Flower	(C) Polygonaceae
(4) Cup and Saucer plant	(D) Convolvulaceae
(5) Wood Rose	(E) Bignoniaceae

(1) C, (2) E, (3) A, (4) B, (5) D

(1) Twiner	(A) Antigonon
(2) Rambler	(B) Tecoma
(3) Creeper	(C) Lonicera
(4) Climber	(D) Antirhinum
(5) Annual	(E) Quisqualis

(1) C, (2) E, (3) B, (4) A, (5) D

(1) *Areca lutescens*	(A) Silver date palm
(2) *Oreodoxa regia*	(B) Areca palm
(3) *Phoenix roebelenii*	(C) Royal palm
(4) *Caryota urens*	(D) Miniature date palm
(5) *Phoenix sylvestris*	(E) Fish tail palm

(1) B, (2) C, (3) D, (4) E, (5) A

(1) Hen and Chicken fern	(A) *Adiantum capillus*
(2) Birds nest fern	(B) *Pteris cretica*
(3) Table fern	(C) *Alsophila crinita*
(4) Tree fern	(D) *Asplenium sp.*
(5) Venus hair fern	(E) *Adiantum bulbiferum*

(1) E, (2) D, (3) B, (4) C, (5) A

(1) Indian Rose wood	(A) *Epipremnum aureum*
(2) Silver ever green	(B) *Aspidistra elatior*
(3) Cast iron plant	(C) *Chamaedorea elegans*
(4) Golden pothos	(D) *Agleonima commutatum*
(5) Parlour palm	(E) *Dalbergia sisoo*

(1) E, (2) D, (3) B, (4) A, (5) C

(1) Kentucky blue grass	(A) *Agrostis palustris*
(2) Red top	(B) *Poa compressa*
(3) Cripping bent grass	(C) *Phleum pratensea*
(4) Canada blue grass	(D) *Poa pratensis*
(5) Timothy	(E) *Agrostis alba*

(1) D, (2) E, (3) A, (4) B, (5) C

(1) Crested wheat grass	(A) *Agrostis tenuis*
(2) Alkali grass	(B) *Fescuta rubra*
(3) Colonial bent grass	(C) *Poa annua*
(4) Annual blue grass	(D) *Agropyron cristatuam*
(5) Creeping red fescue	(E) *Puccinnellia distans*

(1) D, (2) E, (3) A, (4) C, (5) B

(1) Bahia grass	(A) *Ammophila spp.*
(2) Centipede grass	(B) *Paspalum notatum*
(3) Beach grass	(C) *Eremochloa ophiuroides*
(4) Grama grass	(D) *Cynodon dectylon*
(5) Barmuda grass	(E) *Bouteloua spp.*

(1) B, (2) C, (3) A, (4) E, (5) D

(1) Japanese lawn grass	(A) *Zoysia matrella*
(2) Buffalo grass	(B) *Axonopus affinis*
(3) Manilla grass	(C) *Zoysia japonica*
(4) Carpet grass	(D) *Dactylis glomerata*
(5) Orchard grass	(E) *Buchloe dactyloides*

(1) C, (2) E, (3) A, (4) B, (5) D

(1) Rye grass	(A) *Bromus inermis*
(2) Valvet bent grass	(B) *Stenophrum secundatum*
(3) Smooth brome grass	(C) *Agrostis canina*
(4) St. Augustine grass	(D) *Chloris gayana*
(5) Rhodes grass	(E) *Lolium spp.*

(1) E, (2) C, (3) A, (4) B, (5) D

(1) William Robinson	(A) French gardens
(2) William Kent	(B) English gardens
(3) Percy Lancaster	(C) Wild garden
(4) Le Corbuiser	(D) Chandigarh city
(5) Le Notre	(E) Agri-horticultural Society gardens

(1) C, (2) E, (3) B, (4) D, (5) A

(1) *Kigeria Africana*	(A) Copper pod
(2) *Madhuca indica*	(B) African tulip tree
(3) *Magnolia grandiflora*	(C) African sausage tree
(4) *Peltophorum pterocarpum*	(D) Butter tree
(5) *Spathodia campanulata*	(E) Lily tree

(1) C, (2) D, (3) E, (4) A, (5) B

(1) Carnation	(A) In-vitro seed culture
(2) Bougainvillea	(B) Offsets
(3) Orchid	(C) Cane cutting
(4) Agave	(D) Terminal shoot cutting
(5) Dracaena	(E) Hard wood cutting

(1) D, (2) E, (3) A, (4) B, (5) C

(1) Nurjahan	(A) Damascene
(2) Mohini	(B) Polyanthas
(3) Himroz	(C) Rose attar
(4) Swati	(D) Miniature
(5) Pixie	(E) Floribunda

(1) C, (2) E, (3) A, (4) B, (5) D

(1) *Rosa bourboniana*	(A) Dog rose
(2) *Rosa moschata*	(B) Yellow rose
(3) *Rosa canina*	(C) Edward rose
(4) *Rosa foetida*	(D) Cabbage rose
(5) *Rosa centifolia*	(E) Musk rose

(1) C, (2) E, (3) A, (4) B, (5) D

(1) Thimma	(A) Portulaca
(2) Motia	(B) Coreopsis
(3) Pusa Basanti	(C) Bougainvillea
(4) Pusa Tara	(D) Marigold
(5) Jhumka	(E) Jasmine

(1) C, (2) E, (3) D, (4) B, (5) A

(1) Suvasini	(A) Jasmine
(2) Manali	(B) Sweet pea
(3) Surabhi	(C) Amaryllis
(4) Spencer	(D) Tuberose
(5) Chitwan	(E) Dahlia

(1) D, (2) E, (3) A, (4) B, (5) C

(1) *Tagetes patula*	(A) Fox globe
(2) *Acroclinum roseum*	(B) African marigold
(3) *Digitalis purpurea*	(C) Cone flower
(4) *Tagetes erecta*	(D) Paper flower
(5) *Rudbechia bicolor*	(E) French marigold

(1) E, (2) D, (3) A, (4) B, (5) C

(1) Sweet pea	(A) *Salvia splendens*
(2) Straw flower	(B) *Limonium sinuatum*
(3) Sage	(C) *Lathyrus odoratus*
(4) Statice	(D) *Bellis perenis*
(5) Daisy	(E) *Helichrysum bracteatum*

(1) C, (2) E, (3) A, (4) B, (5) D

(1) Begonia	(A) Iridaceae
(2) Canna	(B) Brassicaceae
(3) Crocus	(C) Nymphaeaceae
(4) Nelumbo	(D) Begoniaceae
(5) Stock	(E) Cannaceae

(1) D, (2) E, (3) A, (4) C, (5) B

(1) Celosia	(A) Lamiaceae
(2) Pansy	(B) Zingiberaceae
(3) Salvia	(C) Amaranthaceae
(4) Alpinia	(D) Alliaceae
(5) Allium	(E) Violaceae

(1) C, (2) E, (3) A, (4) B, (5) D

(1) Gladiolus	(A) Shrub
(2) Jacaranda	(B) Summer annual
(3) Azalea	(C) Corm
(4) Kochia	(D) Winter annual
(5) Lobelia	(E) Tree

(1) C, (2) E, (3) A, (4) B, (5) D

(1) Coleus	(A) Indoor plant
(2) Zoysia	(B) Bulbil
(3) Aspidistra	(C) Coloured foliage
(4) Caladium	(D) Lawn grass
(5) Tiger lily	(E) Tuber

(1) C, (2) D, (3) A, (4) E, (5) B

(1) Summer flowering tree	(A) *Bauhinia purpurea*
(2) Spring flowering tree	(B) Silk cotton tree
(3) Winter flowering tree	(C) Gulmohar
(4) Autumn flowering tree	(D) Koelreuteria
(5) Tree valued for fruits	(E) Palas

(1) C, (2) E, (3) B, (4) A, (5) D

(1) Winter flowering shrub	(A) Golden shower
(2) Autumn flowering shrub	(B) Poinsettia
(3) Winter flowering climber	(C) Azalea
(4) Rainy season annual	(D) *Cassia glauca*
(5) Shrub with bracts	(E) Gaillardia

(1) C, (2) D, (3) A, (4) E, (5) B

(1) Acroclinum	(A) South Africa
(2) Ageratum	(B) India
(3) English daisy	(C) Mexico
(4) Cosmos	(D) Australia
(5) Gomphrena	(E) Europe

(1) D, (2) A, (3) E, (4) C, (5) B

(1) Sweet pea	(A) Mediterranean region
(2) Digitalis	(B) Tropical Asia
(3) Celosia	(C) Caucasus
(4) Gypsophila	(D) Western Europe
(5) Stock	(E) Sicily

(1) E, (2) D, (3) B, (4) C, (5) A

(1) Kadamb	(A) *Azadirachta indica*
(2) Araucaria	(B) *Bauhinia variegata*
(3) Neem	(C) *Barringtonia speciosa*
(4) Indian oak	(D) *Neolamrackia cadamba*
(5) Kachnar	(E) *Araucaria columnaris*

(1) D, (2) E, (3) A, (4) C, (5) B

(1) *C. grandis*	(A) Red Cassia
(2) *C. javanica*	(B) Burmese Cassia
(3) *C. marginata*	(C) Pink Cassia
(4) *C. nodosa*	(D) Java Cassia
(5) *C. renigera*	(E) Horse Cassia

(1) E, (2) D, (3) A, (4) C, (5) B

(1) Devil's tree	(A) Fabaceae
(2) Arucaria	(B) Casuarinaceae
(3) White Kachnar	(C) Apocynaceae
(4) Beef wood tree	(D) Araucariaceae
(5) Amaltas	(E) Fabaceae

(1) C, (2) D, (3) E, (4) B, (5) A

(1) Gul-mohar	(A) March-May
(2) Blue gum	(B) April-June
(3) White Gulmohar	(C) April-June
(4) Indian coral tree	(D) March-April
(5) Silver oak	(E) Feb-March

(1) C, (2) D, (3) E, (4) A, (5) B

(1) *Hibiscus tiliaceus*	(A) Chilkhan
(2) *Ginkgo biloba*	(B) Cotton tree
(3) *Filicium decipiens*	(C) Golden bottle brush
(4) *Ficus retusa*	(D) Fossil tree
(5) *Melaleuca bracteata*	(E) Fern tree

(1) B, (2) D, (3) E, (4) A, (5) C

(1) Pride of India	(A) April-July
(2) Lagerstroemia	(B) March-April
(3) Sausage tree	(C) March-April
(4) Anjan	(D) April-Aug
(5) Tree of life	(E) July-Sept

(1) D, (2) E, (3) A, (4) B, (5) C

(1) Maulsari	(A) Annonaceae
(2) Champa	(B) Salicaceae
(3) Chenar	(C) Sapotaceae
(4) Ashok tree	(D) Magnoliaceae
(5) Poplar	(E) Platanaceae

(1) C, (2) D, (3) E, (4) A, (5) B

(1) *Terminalia arjuna*	(A) Jalpatri
(2) *Terminalia cattapa*	(B) Lahura
(3) *Salix babylonica*	(C) Arjuna tree
(4) *Putranjiva roxburghii*	(D) Desi badam
(5) *Tecomella undulata*	(E) Weeping willow

(1) C, (2) D, (3) E, (4) A, (5) B

(1) *C. nodosa*	(A) Yellow
(2) *Hibiscus tiliaceus*	(B) Deep red
(3) *Tecomella undulata*	(C) Pink
(4) *Sterculia villosa*	(D) Light yellow
(5) *Rhododendron arboretum*	(E) Yellow-orange

(1) C, (2) D, (3) E, (4) A, (5) B

(1) Standard	(A) 8×7cm to 28×23cm
(2) Obake	(B) 10×6cm
(3) Tulip	(C) 2×11cm to 20x18cm
(4) *Anthurium* spacing	(D) 3-4 leaf stage
(5) Transplanting	(E) 45 × 45cm

(1) C, (2) A, (3) B, (4) E, (5) D

(1) June-grass	(A) *Cynodon dactylon*
(2) Doob	(B) *Cyperus rotundus*
(3) Motha	(C) *Poa pratensis*

(1) C, (2) A, (3) B

(1) Rajat Rekha	(A) Golden yellow steaks
(2) Swarna Rekha	(B) Silvery white streak
(3) Shringar	(C) Shringar' x 'Mexican Single
(4) Suvasini	(D) Single type
(5) Prajwal	(E) Double type
(6) Vaibhav	(F) Semi-double

(1) B, (2) A, (3) D, (4) E, (5) C, (6) F

(1) Water lilies	(A) *Nelumbo spp.*
(2) Duckweed	(B) *Eichhornia crassipes*
(3) Water fern	(C) *Nymphaea spp*
(4) Water hyacinth	(D) *Lemna sp*
(5) Lotus	(E) *Salvinia minima*

(1) C, (2) D, (3) E, (4) B, (5) A

(1) Ferns	(A) Digitalis
(2) Foliage plants	(B) Hydrangea
(3) Cut flower	(C) Alocasia
(4) Perennials	(D) hyacinth
(5) Bulbs	(E) Pteris

(1) D, (2) C, (3) A, (4) B, (5) E

(1) Albizzia lebbek	(A) Feb-March
(2) Bauhinia variegate	(B) Year round
(3) Bauhinia purpurea	(C) Feb-May
(4) Cassia fistula	(D) April-May
(5) Cordia sebestina	(E) October-March

(1) D, (2) A, (3) E, (4) C, (5) B

(1) Bird's nest fern	(A) *A. bulbiferum*
(2) Table fern	(B) *Alsophila crinita*
(3) Venus hair fern	(C) *Asplenium sp.*
(4) Tree fern	(D) *Adiantum capillus*
(5) Hen and chicken fern	(E) *Pteris cretica*

(1) C, (2) E, (3) D, (4) B, (5) A

(1) Hibiscus	(A) Vishnu Swarup
(2) Marigold	(B) Swami Vinayananda
(3) Chrysanthemum	(C) Bhupal Rao
(4) Jasmine	(D) R.N. Bhatt
(5) Dahlia	(E) SK Datta

(1) D, (2) A, (3) E, (4) C, (5) B